Periodic Table of the Elements

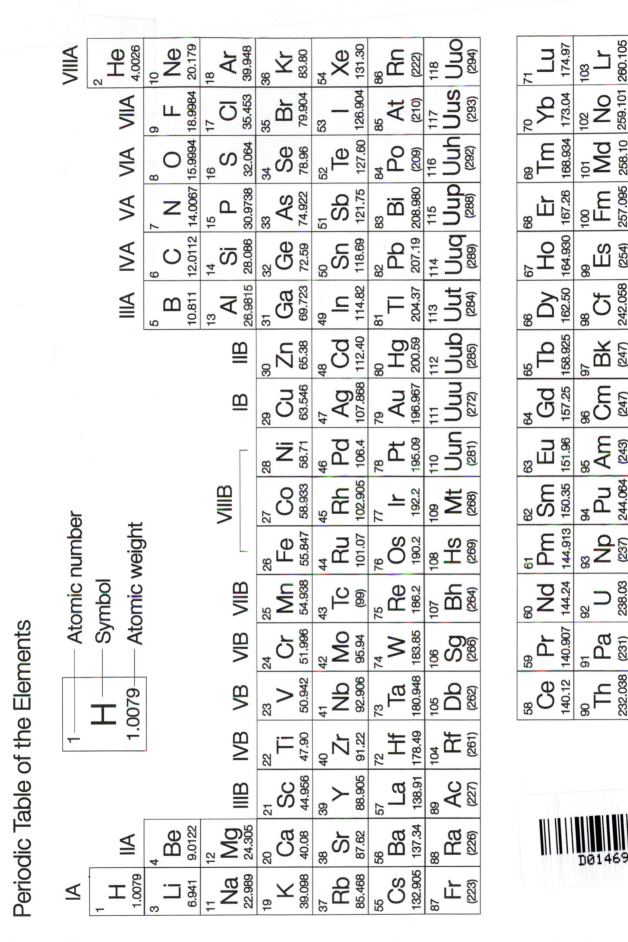

Laboratory Manual

to accompany

Physical Science

Twelfth Edition

Prepared by
Bill W. Tillery
Arizona State University

LABORATORY MANUAL TO ACCOMPANY

PHYSICAL SCIENCE, TWELFTH EDITION

1 2 3 4 5 6 QVS 23 22 21 20 19

ISBN 978-1-260-41131-7
MHID 1-260-41131-1

Cover image: ©Shutterstock/D K Grove

mheducation.com/highered

Contents

Appendix .... **Page**

Introduction

This is a laboratory guide for a one-year college physical science course for nonscience majors. Laboratory sessions described by the guide are designed to provide a hands-on introduction to experimental methods of scientific investigation. Each provides opportunities for you to learn the practical knowledge necessary for a well-rounded understanding of physical science. You will learn the theory and scientific laws pertaining to physical phenomena in the lecture portion of the course. Conducting experiments and collecting data to test the validity of theories and laws requires a different set of skills than those required for success in the lecture. Success in the laboratory involves skills in making accurate measurements of physical quantities in the real world, then formulating valid generalizations and principles based on the data. Every experiment in the laboratory will provide lessons and opportunities to learn these skills. The open-ended *Invitations to Inquiry* can be used alone, as stand-alone investigations, or as special projects along with the more structured learning environment. In addition, there is a *Special Project* (Lab 51) designed to serve as a guide for a longer term, open ended student investigation.

The instructions for each experiment in a structured learning environment include some basic theory and relationships about the physical phenomena to be investigated, data tables, graph paper, and questions to be answered. These are not "fill in the blank" type questions as are typically found in a laboratory workbook, but are designed to help students make a thoughtful analysis with careful, thorough thinking. The instructions for inquiry experiments, on the other hand, are more toward suggestions for interesting, challenging investigations. Less student support is provided for the inquiry experiments, which are usually more successfully completed by more mature students.

For the student. Always bring your calculator and this laboratory guide to each session, completing required calculations and your laboratory report as you work. Be as accurate and neat as possible but do not waste time on a report. Sloppy work is to be avoided, of course, but *concentrate on knowing what you are doing, gathering accurate data, solving problems, and working out conclusions* while in the laboratory. These are the criteria for evaluating laboratory reports, not how much time you spend detailing a handsome report. Focus on these criteria during each laboratory session. In physics experiments, this means accomplishing the purpose within a reasonable error, reasoning what constitutes a reasonable error, and stating what the origin of this error might be. In most of the labs, the purpose is to verify a law or theory that you have covered in class. Therefore, the conclusion may be as simple as "Newton's third law of motion was verified within 10% of the expected value." This should be followed by a statement as to why 10% is an acceptable error for this particular experiment. The conclusions should be reasonable and make sense, not necessarily agreeing with "expected" findings. Thoughtful analysis and careful, thorough thinking are much more important (and reasonable) than 0% error. In chemistry experiments, the purpose of a lab is often to understand how matter interacts and the procedures and tests used in understanding matter. The results of procedures and tests are often used to analyze and reason, and to identify unknowns. Analysis of findings and possible errors are appropriate in these laboratory experiments as well as in the earth science experiments.

You are encouraged to work together in study groups, but your work should be your own. Note all the appendices at the end of this guide. They cover most of the problems that past students have had with laboratory skills and procedures.

Acknowledgments

This revision of the *Laboratory Manual in Physical Science* is based on the critical reviews and comments of the many users and reviewers of its previous editions. The author wishes to thank them all, acknowledging their significant contributions. The author is also grateful to reviewers of the 4th edition and their valuable contributions: Kyle L. Hathcox, Union University; Gene Hopkins, Rogers State University; Wilda M. Pounds, Northeast MS Community College; and, Cu G. Phung, Methodist College. As always, the author welcomes comments and suggestions from instructors and students. Comments and suggestions about the *Laboratory Manual in Physical Science* or any other part of the program should be addressed to Bill W. Tillery at the email address tillery@asu.edu.

The astronomy investigations of Celestial Coordinates, Motions of the Sun, Phases of the Moon, and Power Output of the Sun are from the astronomy manual of *Laboratory Astronomy: Experiments and Exercises*, Anthony J. Nicastro, Wm. C. Brown Publishers, 1990. The geology investigation of Topographic Maps is partly from the *Laboratory Manual for Physical Geology*, Norris W. Jones and Charles E. Jones, Mcgraw-Hill Higher Education, 2003, pp. 90–91.

Materials Required for Each Experiment
(Quantities given are for an individual or teams of students.)

1. **Graphing** (page 1): Meterstick, masking tape, several types of small balls that bounce (tennis, ping-pong, rubber hand ball, etc.).

2. **Ratios** (page 15): Several sizes of cups or beakers with round bottoms for tracing circles, metric rulers, string, wood or plastic cubes (about 30), three sizes of rectangular containers to hold water, balance, graduated cylinder.

3. **Motion** (page 27): Battery-operated toy bulldozer (or other toy car), long sheets of computer paper, butcher paper, or adding machine paper; masking tape, meterstick, stopwatch, inclined ramp (1 m or longer), 1 to 2 m rolling ball track, steel ball or marble (alternate setup: lab cart, photogates, computer with timer software).

4. **Free Fall** (page 43): Free-fall apparatus with spark timer, mass, meterstick, metric ruler (alternate setup: lab cart on track, photogate and pulley with spokes, computer with timer software).

5. **The Pendulum** (page 53): Different masses for bobs, support for pendulum, light cord or nylon, meterstick, stopwatch.

6. **Projectile Motion** (page 65): Ramp for table top, ring stand and ring, small rubber ball, meterstick, stop watch, Styrofoam cup.

7. **Newton's Second Law** (page 75): Air track with cart, spark timer (alternate setup: lab cart on track, photogate and pulley with spokes, computer with timer software), pulley, nylon string, mass hanger, five 50 g masses, metric ruler or meterstick.

8. **Conservation of Momentum** (page 87): Air track with two carts, spark timer or stop watch (alternate setup: lab cart on track, photogate and pulley with spokes, computer with timer software), mass set, metric ruler or meterstick.

9. **Rotational Equilibrium** (page 95): Meterstick with knife-edge clamp and support stand, three movable mass hangers and meterstick clamps, mass set.

10. **Centripetal Force** (page 105): Mass hanger and small masses, nylon string, rubber stopper, holder (wood or plastic rod with small hole for string; see figure 10.1).

11. **Archimedes' Principle** (page 111): Laboratory balance, object such as large metal mass, overflow can and catch bucket, spring scale (or laboratory balance attached to ring stand), block of wood, ball of clay.

12. **Boyle's Law** (page 117): Boyle's law apparatus, barometer.

13. **Work and Power** (page 125): Stairwell, meterstick, stopwatch, and a scale for weighing people.

14. **Friction** (page 129): Laboratory balance, wood block, clean and dry board, pulley and support, light cord, weight hanger and masses, set of laboratory masses, including a 0.5 kg mass.

15. **Hooke's Law** (page 139): Laboratory spring, support for spring, weight hanger, masses for weight hanger, meterstick.

16. **Thermometer Fixed Points** (page 147): Laboratory thermometer, beakers, crushed ice, steam generator, laboratory mercury barometer.

17. **Absolute Zero** (page 155): Apparatus (see Fig. 17.1 on page 156) of large glass tube fitted with 2-hole rubber stoppers, small glass tube longer than large tube, two short glass tubes, rubber tubing and clamp, meterstick, mercury, thermometer, large beaker, funnel, crushed ice, hot water, steam generator.

18. **Specific Heat** (page 165): Two Styrofoam cups to serve as a calorimeter, balance, three samples of shot made of different metals (e.g., aluminum, copper, lead), heating source for boiling water (hot plate and beaker or burner and ring stand setup), thermometer.

19. **Static Electricity** (page 173): Two glass rods, two hard rubber rods, nylon or silk cloth, wool cloth or fur, thread, electroscope, two rubber balloons.

20. **Electric Circuits** (page 179): Two flashlight batteries, tape or two flashlight battery holders, two flashlight bulbs, two flashlight bulb holders, hookup wire.

21. **Series and Parallel Circuits** (page 185): Three #41 bulbs (0.5 A), bulb sockets, two dry cells (or other source of 3 V dc), a dc voltmeter (0 to 5 volt range), and a dc ammeter (0 to 2 amp range), switch, hookup wire or patch cords.

22. **Ohm's Law** (page 193): Adjustable dc power supply, dc voltmeter, dc ammeter, resistors, hookup wire or patch cords.

23. **Magnetic Fields** (page 203): Large sheet of unlined white paper, small magnetic compass, bar magnet, sharp pencil, large plastic sheet or glass plate, iron filings.

24. **Electromagnets** (page 207): Galvanoscope, small magnetic compass, dry cell or laboratory source of 1.5 V dc, about 3 m of No. 18 copper wire, 1/2 cm diameter soft iron spike about 12 cm long, box of regular-size paper clips.

25. **Standing Waves** (page 211): Nylon string of known density, vibrator, pulley, mass hanger and small masses, meterstick, adjustable stroboscope (optional).

26. **Speed of Sound** (page 219): Resonance tube apparatus, meterstick, two tuning forks of different pitches, rubber hammer, thermometer.

27. **Reflection and Refraction** (page 227): Ruler, cardboard (from box), small flat mirror, small wood block, rubber bands, straight pins, unlined white paper, protractor, 5 cm square glass plate.

28. **Physical and Chemical Change** (page 233): Sodium chloride, graduated cylinder, evaporating dish, nichrome wire, tongs, magnesium ribbon, small test tube, dilute hydrochloric acid, silver nitrate solution, funnel, funnel support, ring stand, filter paper, copper(II) chloride solution, beaker, aluminum foil.

29. **Hydrogen** (page 239): About 20 g mossy zinc, dilute sulfuric acid (1 vol. concentrated H_2SO_4 to 4 vol. water), long wood splints, apparatus shown in figure 20.1, three small collecting bottles (250 mL or less), three glass plates, copper(II) sulfate solution, small test tube.

30. **Oxygen** (page 245): Mixture of two parts of potassium chlorate ($KClO_3$) and one part of manganese dioxide (MnO_2), large Pyrex test tube and other apparatus shown in figure 19.1, four 500 mL collecting bottles, four glass plates, wood splints, limewater, charcoal stick, forceps, steel wool.

31. **Conductivity of Solutions** (page 251): Conductivity apparatus (figure 21.1), short patch cord with alligator clips, distilled water, tap water, dry sodium chloride (table salt), sugar solution (1%), ethyl alcohol, hydrochloric acid (1.0 M: 81 mL concentrated/L solution), sodium hydroxide (1.0 M: 40 g solid/L solution), sodium chloride solution (1.0 M: 58 g solid/L solution), vinegar, glycerin.

32. **Percentage Composition** (page 257): Test tube (to be discarded after the experiment), test tube holder, burner, balance, desiccator (or large jar with lid partly filled with anhydrous calcium chloride), sucrose (table sugar).

33. **Metal Replacement Reactions** (page 263): Metal strips (about 2 cm × 5 cm) of copper, zinc, and lead; sandpaper, test tubes, test tube rack, graduated cylinder, 10 cm length of thin copper wire, silver nitrate solution (0.1 M: 17.0 g/L of solution), copper nitrate solution (0.1 M: 24.2 g/L of solution), lead nitrate solution (0.1 M: 33.1 g/L of solution).

34. **Producing Salts by Neutralization** (page 269): Evaporating dish, small beakers, dropper, phenolphthalein solution, graduated cylinder, glass stirring rod, universal indicator paper (with color coded pH scale), watch glass, ring stand and ring, wire screen, burner, hydrochloric acid (1.0 M: 81 mL concentrated/L of solution), sulfuric acid (1.0 M: 56 mL concentrated/L of solution), sodium hydroxide solution (1.0 M: 40 g/L of solution), potassium hydroxide solution (1.0 M: 56 g/L of solution), distilled water.

35. **Identifying Salts** (page 275): Beakers, burner, medicine dropper, cobalt glass squares, platinum or nichrome wire, forceps, test tubes and rack, graduated cylinder, pipette (or thin glass tube). Flame test solutions: sodium nitrate solution (1.0 M: 85 g/L of solution), lithium nitrate (1.0 M: 123 g/L of solution), strontium nitrate (1.0 M: 284 g/L of solution), calcium nitrate (1.0 M 163 g/L of solution), barium nitrate (1.0 M: 261 g/L of solution), potassium nitrate (1.0 M: 101 g/L of solution), dilute hydrochloric acid (1:4), and distilled water. Chemical test solutions: silver nitrate solution (0.1 M: 17.3 g/L of solution), barium chloride solution (0.1 M: 24.4 g/L of solution), calcium carbonate solution (saturated), iron(II) sulfate solution (saturated), concentrated sulfuric acid, sodium chloride solution (0.1 M: 5.9 g/L of solution), and potassium sulfate solution (0.1 M: 15.8 g/L of solution).

36. **Solubility Curves** (page 281): Ring stand and ring, burner, beakers, evaporating dish, wire screen, balance, graduated cylinder, distilled water, crushed ice, potassium dichromate, thermometer, glass stirring rod, spatula, laboratory forceps.

37. **Natural Water** (page 289): Distilled water, 3% alum solution, aqueous ammonium hydroxide (6.0 M), beakers, dropper, glass stirring rod, filtering funnel, ring stand and ring, glass wool, clean sand, test tubes and rack, wire screen, burner, watch glass, graduated cylinder, calcium chloride solution (0.001 M: 0.1 g/L of solution), standard soap solution, muddy water, tap or well water.

38. **Measurement of pH** (page 297): Solutions of hydrochloric acid (0.1 M: 8.1 mL concentrated/L of solution), sulfuric acid (0.1 M: 5.6 mL concentrated/L of solution), acetic acid (0.1 M: 5.6 mL concentrated/L of solution), sodium hydroxide (0.1 m: 4 g/L of solution), barium hydroxide (17 g/L of solution), household ammonia (diluted with 6 volumes water). Indicators: red litmus paper, blue litmus paper, bromthymol blue, methyl orange, methyl red, phenolphthalein, and universal indicator paper (with color-coded pH scale). Watch glass, glass stirring rod, test tubes and rack. Various unknown solutions (different acid or base solutions of different concentrations).

39. **Amount of Water Vapor in the Air** (page 305): Sling psychrometer (or two thermometers with a 3 cm length of cotton shoelace on the bulb end of one), meterstick.

40. **Nuclear Radiation** (page 311): Geiger counter, radioactive source (gamma emitting), meterstick, sheets of lead foil, two ring stands and ring stand clamps, alpha and beta emitting radioactive sources (optional), materials with various attenuation properties (optional).

41. **Growing Crystals** (page 323): Large Pyrex beaker (1 liter or larger), selected salt or salts (see table 26.1 for possibilities and approximate amounts of salt required for each), ring stand and ring, wire screen, burner, balance, glass stirring rod, glass jar (1 liter or larger with lid), spatula, very fine fishing leader, tape, epoxy glue.

42. **Properties of Common Minerals** (page 329): Mineral collection, streak plate, magnifying glass, steel file, pocketknife, copper penny, magnet, dilute hydrochloric acid, dropper, balance, graduated cylinder and overflow can to determine density (optional).

43. **Density of Igneous Rocks** (page 335): Balance, overflow can, graduated cylinder, ring stand and ring, thin nylon string, beaker, granite specimen, basalt specimen.

Experiment 1: Graphing

Introduction

The purpose of this introductory laboratory exercise is to gain experience in gathering and displaying data from a simple experiment. Refer to figure 1.1 for terminology used when discussing a graph and see appendix I for a detailed discussion about the terms.

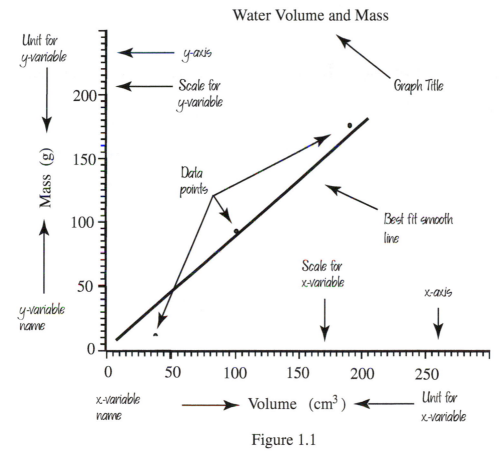

Figure 1.1

Procedure

1. Position a meterstick vertically on a flat surface, such as a wall or the side of a lab bench. Be sure the metric scale of the meterstick is on the outside and secure the meterstick to the wall or lab bench with two strips of masking tape.

2. Drop a ball as close as possible to the meterstick and measure (a) the height dropped, and (b) the resulting height bounced. Repeat this for five different heights dropped and record all data in Data Table 1.1 on page 6. In the data table, identify the *independent* (manipulated) variable and the *dependent* (responding) variable.

1

3. Use the graph paper on page 9 to make a graph of the data in Data Table 1.1, being sure to follow all the rules of graphing (see appendix I on page 395 for help). Title the graph, "Single Measurement Bounce Height as a Function of Height Dropped."

4. After constructing the graph, but before continuing with this laboratory investigation, answer the following questions:

    (a) What decisions did you have to make about how you conducted the ball-dropping investigation?

    (b) Would you obtain the exact same result if you dropped the ball from the same height several times? Explain.

    (c) Did you make a dot–to–dot line connecting the data points on your graph? Why or why not?

    (d) Could you use your graph to obtain a predictable result for dropping the ball from different heights? Explain why you could or could not.

(e) What is the significance of the origin on the graph of this data? Did you use the origin as a data point? Why or why not?

5. Make five more measurements for each of the previous five *height-dropped* levels. Find the *average height bounced* for each level and record the data and the average values in Data Table 1.2 on page 6.

6. Make a new graph of the *average height bounced* for each level that the ball was dropped. Draw a *straight best fit line* that *includes the origin* by considering the general trend of the data points. Draw the straight line as close as possible to as many data points as you can. Try to have about the same number of data points on both sides of the straight line. Title this graph, "Averaged Bounce Height."

7. Compare how well both graphs, "Single Measurement Bounce Height" and "Averaged Bounce Height," predict the heights that the ball will bounce for *heights dropped* that were not tried previously. Locate an untried height-dropped distance on the straight line, then use the corresponding value on the scale for height bounced as a prediction. Test predictions by noting several different heights, then measuring the actual heights bounced. Record your predictions and the actual experimental results in Data Table 1.3 on page 7.

8. Use a new, *different kind of ball* and investigate the bounce of this different ball. Record your single-measurement data for this different ball in Data Table 1.4. Record the averaged data for the height of the bounce for the five levels of dropping in Data Table 1.5. Repeat procedure step 7 for the different kind of ball. Record your predictions and the actual results in Data Table 1.6 on page 8.

9. Graph the results of the different kind of ball investigations onto the two previous graphs. Be sure to distinguish between sets of data points and lines by using different kinds of marks. Explain the meaning of the different marks in a *key* on the graph.

10. What does the steepness (slope) of the lines tell you about the bounce of the different balls?

Results

1. Describe the possible sources of error in this experiment.

2. Describe at least one way that data concerning two variables is modified to reduce errors in order to show general trends or patterns.

3. How is a graph modified to show the best approximation of theoretical, error-free relationships between two variables?

4. Compare the usefulness of a graph showing (a) exact, precise data points connected dot–to–dot and (b) an approximated straight line that has about the same number of data points on both sides of the line.

5. Was the purpose of this lab accomplished? Why or why not? (Your answer to this question should be reasonable and make sense, showing thoughtful analysis and careful, thorough thinking.)

Invitation to Inquiry

Measurements of variables involved in some physical relationship will often yield a straight line, and this relationship is said to be direct, or *linear*. There are more types of relationships between variables, and most can be identified as producing one of five basic shapes of graphs. These are identified, left to right, as no relationship, linear, inverse, square, and square root.

Data that yields a straight line has a linear relationship and can be described algebraically by the slope-intercept form of an equation, $y = mx + b$ where m is the slope and b is the y intercept. If a graph results in one of the three curves, data can be manipulated to produce a straight-line graph. For example, a graph of the pressure changes of a confined gas that occurs with changes in volume might produce a graph that looks like this:

Since this curve is an inverse relationship, the plot is redone as P vs. 1/V. The graph now looks like this:

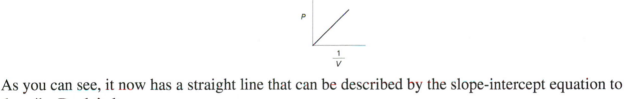

As you can see, it now has a straight line that can be described by the slope-intercept equation to describe Boyle's law.

After giving the possibilities some thought, look for relationships that might result in something other than a direct relationship. Make measurements, graph your data, then decide which of the five shapes the graph resembles. If the data results in one of the curves, try manipulating the data to generate a straight-line graph. For example, obtain a toy dart gun and dart. Vary the weight of the dart by taping small masses to the dart, then measure the height the dart is propelled. What is the relationship between the weight of a dart and the height it is propelled? What is the shape of a graph comparing the weight and height and what does this mean about the relationship?

What other relationships can you find in the lab, outside, or between any two variables in everyday occurrences? Can you find an example of each of the five shapes of graphs?

Data Table 1.1	Single Measurement Data: 1st Ball	
Trial	Height Dropped _____variable	Height Bounced _____variable
1	_____	_____
2	_____	_____
3	_____	_____
4	_____	_____
5	_____	_____

Data Table 1.2	Averaged Bounce Data: 1st Ball					
Dropped Height	Bounce Height					
	Trial 1	Trial 2	Trial 3	Trial 4	Trial 5	Average
_____	_____	_____	_____	_____	_____	_____
_____	_____	_____	_____	_____	_____	_____
_____	_____	_____	_____	_____	_____	_____
_____	_____	_____	_____	_____	_____	_____
_____	_____	_____	_____	_____	_____	_____

Data Table 1.3   Predictions and Results: 1st Ball

Trial	Single Measurement Data			Averaged Data		
	Dropped Height	Predicted Bounce	Measured Bounce	Dropped Height	Predicted Bounce	Measured Bounce
1	_____	_____	_____	_____	_____	_____
2	_____	_____	_____	_____	_____	_____
3	_____	_____	_____	_____	_____	_____

Data Table 1.4   Single Measurement Data: 2nd Ball

Trial	Height Dropped _____variable	Height Bounced _____variable
1	_____	_____
2	_____	_____
3	_____	_____
4	_____	_____
5	_____	_____

Data Table 1.5   Averaged Bounce Data: 2nd Ball

Dropped Height	Bounce Height					
	Trial 1	Trial 2	Trial 3	Trial 4	Trial 5	Average
_____	_____	_____	_____	_____	_____	_____
_____	_____	_____	_____	_____	_____	_____
_____	_____	_____	_____	_____	_____	_____

Data Table 1.6   Predictions and Results: 2nd Ball

Trial	Single Measurement Data			Averaged Data		
	Dropped Height	Predicted Bounce	Measured Bounce	Dropped Height	Predicted Bounce	Measured Bounce
1	_____	_____	_____	_____	_____	_____
2	_____	_____	_____	_____	_____	_____
3	_____	_____	_____	_____	_____	_____

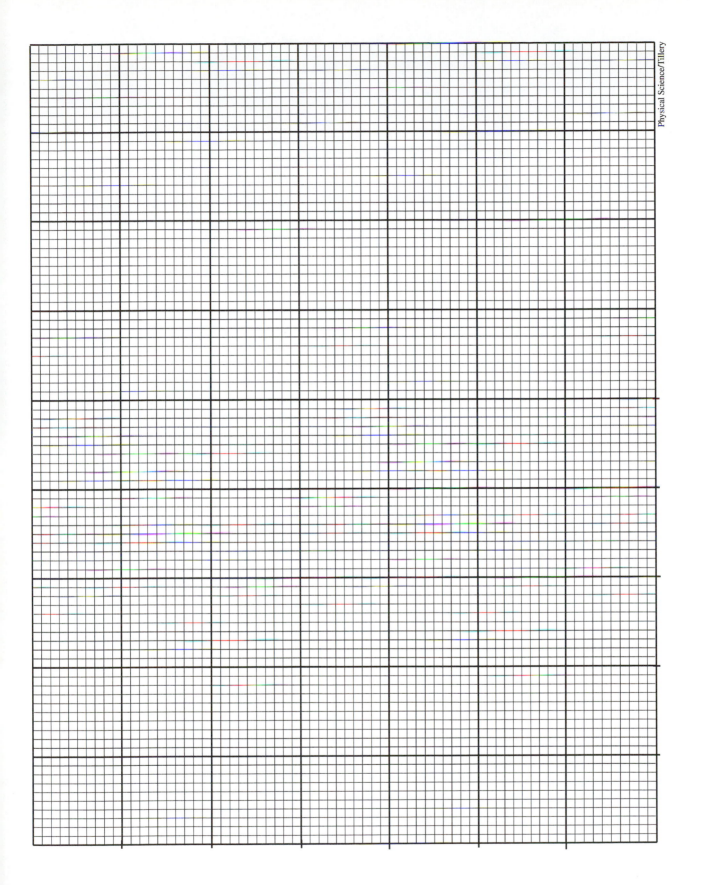

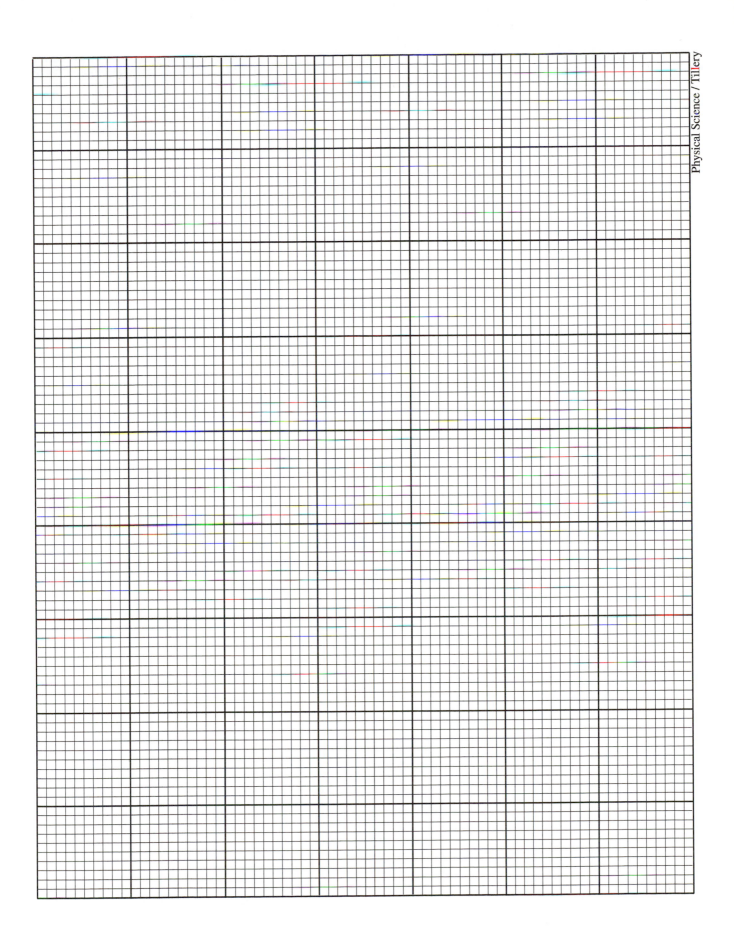

Name_____Section_____Date_____

Experiment 2: Ratios

Introduction

The purpose of this introductory laboratory exercise is to investigate how measurement data are simplified in order to generalize and identify trends in the data. Data concerning two quantities will be compared as a **ratio**, which is generally defined as a relationship between numbers or quantities. A ratio is usually simplified by dividing one number by another.

Procedure

Part A: Circles and Proportionality Constants

1. Obtain three different sizes of cups, containers, or beakers with circular bases. Trace around the bottoms to make three large but different-sized circles on a blank sheet of paper.

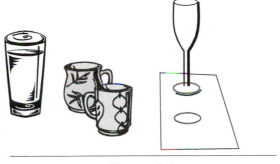

Figure 2.1

2. Mark the diameter on each circle by drawing a straight line across the center. Measure each diameter in mm and record the measurements in Data Table 2.1. Repeat this procedure for each circle for a total of three trials.

3. Measure the circumference of each object by carefully positioning a length of string around the object's base, then grasping the place where the string ends meet. Measure the length in mm and record the measurements for each circle in Data Table 2.1. Repeat the procedure for each circle for a total of three trials. Find the ratio of the circumference of each circle to its diameter. Record the ratio for each trial in Data Table 2.1 on page 23.

4. The ratio of the circumference of a circle to its diameter is known as **pi** (symbol π), which has a value of 3.14... (the periods mean many decimal places). Average all the values of π in Data Table 2.1 and calculate the experimental error.

Part B: Area and Volume Ratios

1. Obtain one cube from the supply of same-sized cubes in the laboratory. Note that a cube has six sides, or six units of surface area. The side of a cube is also called a *face*, so each cube has six identical faces with the same area. The overall surface area of a cube can be found by measuring the length and width of one face (which should have the same value) and then multiplying (length)(width)(number of faces). Use a metric ruler to measure the cube, then calculate the overall surface area and record your finding for this small cube in Data Table 2.2 on page 23.

2. The volume of a cube can be found by multiplying the (length)(width)(height). Measure and calculate the volume of the cube and record your finding for this small cube in Data Table 2.2.

3. Calculate the ratio of surface area to volume and record it in Data Table 2.2.

4. Build a medium-sized cube from eight of the small cubes stacked into one solid cube. Find and record (a) the overall surface area, (b) the volume, and (c) the overall surface area to volume ratio, and record them in Data Table 2.2.

5. Build a large cube from 27 of the small cubes stacked into one solid cube. Again, find and record the overall surface area, volume, and overall surface area to volume ratio and record your findings in Data Table 2.2.

6. Describe a pattern, or generalization, concerning the volume of a cube and its surface area to volume ratio. For example, as the volume of a cube increases, what happens to the surface area to volume ratio? How do these two quantities change together for larger and larger cubes?

Part C: Mass and Volume

1. Obtain at least three straight-sided, rectangular containers. Measure the length, width, and height *inside* the container (you do not want the container material included in the volume). Record these measurements in Data Table 2.3 (page 23) in rows 1, 2, and 3. Calculate and record the volume of each container in row 4 of the data table.

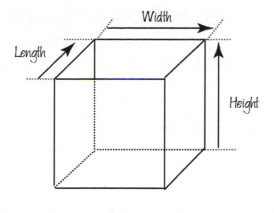

Figure 2.2

2. Measure and record the mass of each container in row 5 of the data table. Measure and record the mass of each container when "level full" of tap water. Record each mass in row 6 of the data table. Calculate and record the mass of the water in each container (mass of container plus water minus mass of empty container, or row 6 minus row 5 for each container). Record the mass of the water in row 7 of the data table.

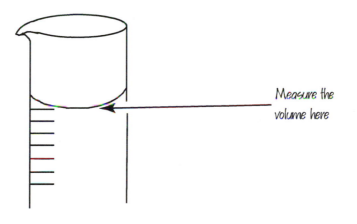

Figure 2.3

3. Use a graduated cylinder to measure the volume of water in each of the three containers. Be sure to get *all* the water into the graduated cylinder. Record the water volume of each container in milliliters (mL) in row 8 of the data table.

4. Calculate the ratio of cubic centimeters (cm^3) to mL for each container by dividing the volume in cubic centimeters (row 4 data) by the volume in milliliters (row 8 data). Record your findings in the data table.

5. Calculate the ratio of mass per unit volume for each container by dividing the mass in grams (row 7 data) by the volume in milliliters (row 8 data). Record your results in the data table.

6. Make a graph of the mass in grams (row 7 data) and the volume in milliliters (row 8 data) to picture the mass per unit volume ratio found in step 5. Put the volume on the *x*-axis (horizontal axis) and the mass on the *y*-axis (the vertical axis). The mass and volume data from each container will be a data point, so there will be a total of three data points.

7. Draw a straight line on your graph that is as close as possible to the three data points and the origin (0, 0) as a fourth point. If you wonder why (0, 0) is also a data point, ask yourself about the mass of a zero volume of water!

8. Calculate the slope of your graph. (See appendix II on page 397 for information on calculating a slope.)

9. Calculate your experimental error. Use 1.0 g/mL (grams per milliliter) as the accepted value.

10. Density is defined as mass per unit volume, or mass/volume. The slope of a straight line is also a ratio, defined as the ratio of the change in the *y*-value per the change in the *x*-value. Discuss why the volume data was placed on the *x*-axis and mass on the *y*-axis and not vice versa.

11. Was the purpose of this lab accomplished? Why or why not? (Your answer to this question should show thoughtful analysis and careful, thorough thinking.)

Results

1. What is a ratio? Give several examples of ratios in everyday use.

2. How is the value of π obtained? Why does π not have units?

3. Describe what happens to the surface area to volume ratio for larger and larger cubes. Predict if this pattern would also be observed for other geometric shapes such as a sphere. Explain the reasoning behind your prediction.

4. Why does crushed ice melt faster than the same amount of ice in a single block?

5. Which contains more potato skins: 10 pounds of small potatoes or 10 pounds of large potatoes? Explain the reasoning behind your answer in terms of this laboratory investigation.

6. Using your own words, explain the meaning of the slope of a straight-line graph. What does it tell you about the two graphed quantities?

7. Explain why a slope of mass/volume of a particular substance also identifies the density of that substance.

Problems

An aluminum block that is 1 m × 2 m × 3 m has a mass of 1.62×10^4 kilograms (kg). The following problems concern this aluminum block:

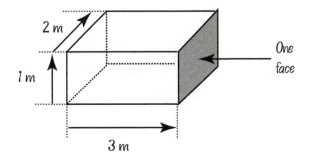

2 m

1 m

3 m

One face

Figure 2.4

1. What is the volume of the block in cubic meters (m^3)?

2. What are the dimensions of the block in centimeters (cm)?

3. Make a sketch of the aluminum block and show the area of each face in square centimeters (cm^2).

4. What is the volume of the block expressed in cubic centimeters (cm^3)?

5. What is the mass of the block expressed in grams (g)?

6. What is the ratio of mass (g) to volume (cm^3) for aluminum?

7. Under what topic would you look in the index of a reference book to check your answer to question 6? Explain.

Invitation to Inquiry

If you have popped a batch of popcorn, you know that a given batch of kernels might pop into big and fluffy popcorn. But another batch might not be big and fluffy and some of the kernels might not pop. Popcorn pops because each kernel contains moisture that vaporizes into steam, expanding rapidly and causing the kernel to explode, or pop. Here are some questions you might want to consider investigating to find out more about popcorn: Does the ratio of water to kernel mass influence the final fluffy size of popped corn? (Hint: measure mass of kernel before and after popping). Is there an optimum ratio of water to kernel mass for making bigger popped kernels? Is the size of the popped kernels influenced by how rapidly or how slowly you heat the kernels? Can you influence the size of popped kernels by drying or adding moisture to the unpopped kernels? Is a different ratio of moisture to kernel mass better for use in a microwave than in a convention corn popper? Perhaps you can think of more questions about popcorn.

Data Table 2.1   Circles and Ratios

	Small Circle			Medium Circle			Large Circle		
Trial	1	2	3	1	2	3	1	2	3
Diameter (D)	___	___	___	___	___	___	___	___	___
Circumference (C)	___	___	___	___	___	___	___	___	___
Ratio of C/D	___	___	___	___	___	___	___	___	___

Average $\dfrac{C}{D}$ = _____   Experimental error: _____

Data Table 2.2   Area and Volume Ratios

	Small Cube	Medium Cube	Large Cube
Surface Area	_____	_____	_____
Volume	_____	_____	_____
Ratio of Area/Volume	_____	_____	_____

Data Table 2.3	Mass and Volume Ratios		
Container Number	1	2	3
1. Length of container	_____cm	_____cm	_____cm
2. Width of container	_____cm	_____cm	_____cm
3. Height of container	_____cm	_____cm	_____cm
4. Calculated volume	_____cm^3	_____cm^3	_____cm^3
5. Mass of container	_____g	_____g	_____g
6. Mass of container and water	_____g	_____g	_____g
7. Mass of water	_____g	_____g	_____g
8. Measured volume of water	_____mL	_____mL	_____mL
9. Ratio of calculated volume to measured volume of water	_____cm^3/mL	_____cm^3/mL	_____cm^3/mL
10. Ratio of mass of water to measured volume of water	_____g/mL	_____g/mL	_____g/mL

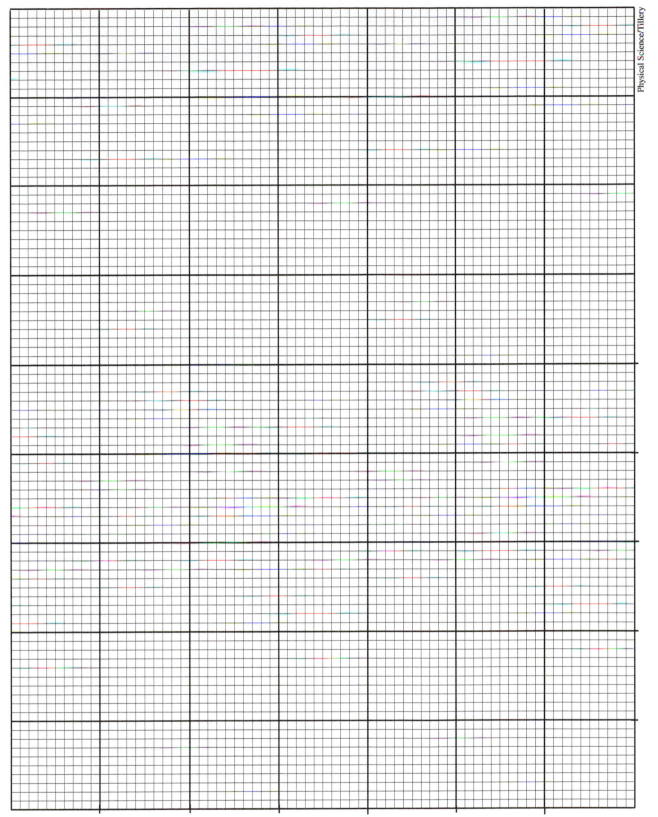

Experiment 3: Motion

Introduction

In this investigation you will analyze and describe motion with a constant velocity and motion with a nonconstant velocity. First, motion with a constant velocity will be investigated by using a battery-operated toy bulldozer, or any toy car or truck that moves at a fairly constant speed. Data will be collected, analyzed, and a concept will be formalized to described what is happening to the toy as it moves.

Figure 3.1 compares the distance vs. time slopes for motion with a constant velocity, with a nonconstant velocity, and with no velocity at all. Note that the slope for some object not moving will be a straight horizontal line. If a vehicle is moving at a uniform (constant) velocity the line will have a positive slope. This slope will describe the magnitude of the velocity, sometimes referred to as the **speed**. The line for a vehicle moving at a nonconstant speed, on the other hand, will be nonconstant as shown in figure 3.1. A nonconstant speed is also known as accelerated motion, and the ratio of how fast the motion is changing per unit of time is called **acceleration**.

Taking measurable data from a multitude of sensory impressions, finding order in the data, then inventing a concept to describe the order are the activities of science. This investigation applies this process to motion.

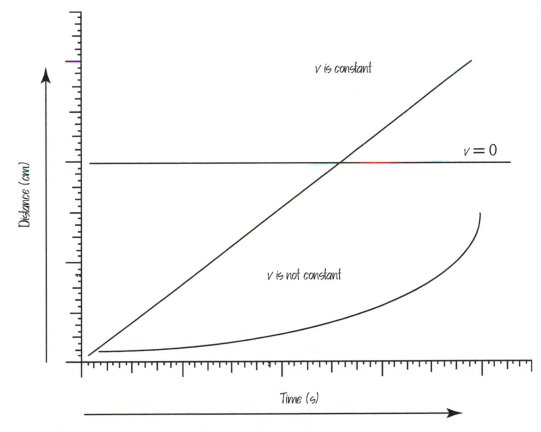

Figure 3.1

Procedure

Part A: Constant Velocity on the Level

1. Use masking tape to secure a length of paper such as long sheets of computer paper, rolled butcher paper, or adding machine tape across the floor. The paper should be long enough so the motorized toy vehicle used will not cross the entire length in less than 8 to 10 seconds. Thus, the exact length of paper selected will depend on the vehicle and battery conditions. (Note: Erratic increases or decreases of speed probably mean that a new battery is needed.) The paper will be used to record successive positions of the toy at specific time intervals.

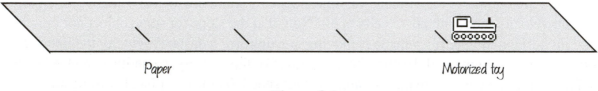

Paper   Motorized toy

Figure 3.2

2. One person with a stopwatch will call out equal time intervals that are manageable but result in at least five or six data points for the total trip. Another person will mark the position of the toy vehicle on the paper when each time interval is called. To avoid interfering with the motion of the toy, mark the position from behind each time. This also means that the starting position should be marked from behind. Other means of measuring velocity that might be used in your laboratory, such as the use of photogates and computer software, will be explained by your instructor.

3. Measure the intervals between the time marks, recording your data in Data Table 3.1 on page 34. Make a graph that describes the motion of the toy vehicle by placing the distance (the dependent variable) on the vertical axis and time (the independent variable) on the horizontal axis. Draw the best straight line as close as possible to the data points. Calculate the slope and record it someplace on the graph.

Part B: Constant Velocity on an Incline

1. This investigation is similar to part A, but this time the toy vehicle will move up an inclined ramp that is at least 1 m long.

2. Elevate the ramp with blocks or books so that 1 meter from the bottom of the ramp is 10 cm high. As in part A, one person with a stopwatch will call out equal time intervals that are manageable, but result in at least five or six data points for the total trip. Another person will mark the position of the toy vehicle on the paper when each time interval is called. To avoid interfering with the motion of the toy, mark the position from behind each time. Also mark the starting position from behind. Measure the intervals between the time marks, recording your data in Data Table 3.2.

3. Elevate the ramp to 20 cm high and repeat procedure step 2.

4. Elevate the ramp to 30 cm high and again repeat procedure step 2. Make a graph of all three sets of data in Data Table 3.2. Calculate the slope of each line and write each somewhere on the graph.

Part C: Motion with Nonconstant Velocity

1. You will now set up a track for collecting data about rolling balls. This track can be anything that serves as a smooth, straight guide for a rolling ball. It could be a board with a V-shaped groove, U-shaped aluminum shelf brackets, or two lengths of pipe taped together, for example.

2. The track should be between 1 and 2 m long and supported somewhere between 10 and 50 cm above the table at the elevated end (figure 3.3). In this investigation, a longer track will mean better results. You should consider 1 m as a *minimum* length. Your instructor will describe a different procedure if your lab has photogates, computer software, or different equipment.

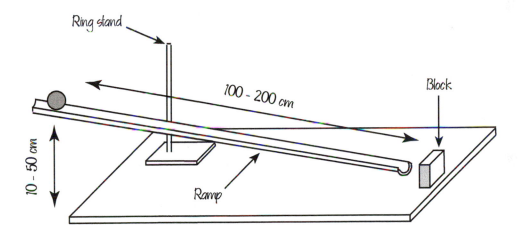

Figure 3.3

3. You will select a minimum of six positions on the ramp from which to release a steel ball or marble. One position should be the uppermost end and the others should be equally spaced. Hold a ruler across the track with the ball behind it, then release the ball by lifting the ruler straight up the same way each time. Start a stopwatch when the ball is released, then stop it when the ball reaches the bottom of the ramp. A block at the bottom of the ramp will stop the ball and the sound of the ball hitting the block will signal when to stop the stopwatch.

4. Measure the distance and time for three data runs, then average the data for each of the six positions. Record the data in Data Table 3.3 on page 36. Make a graph of the data with time on the *x*-axis.

Results

1. Explain how you know a relationship exists between the distance and time variables according to your graphs. How would the graphs appear if there were no relationships?

2. Considering motion with a constant velocity: What happens to changes in distance during equal time intervals? Is this what you would expect?

3. What is the rate of travel of the toy over (a) a flat surface, (b) a surface elevated 10 cm high, (c) a surface elevated 20 cm high, and (d) a surface elevated 30 cm high?

4. Considering motion with a nonconstant velocity: How does the total distance change as the total time increases; that is, do they both increase at the same rate? Explain the meaning of this observation.

5. Again considering motion with a nonconstant velocity: What happens to changes in distance during equal time intervals? Is this what you would expect?

6. Was the purpose of this lab accomplished? Why or why not? (Your answer to this question should show thoughtful analysis and careful, thorough thinking.)

Going Further

In part of this investigation, you learned that $\bar{v} = \dfrac{d}{t}$ Using this equation, explain how you can find

(a) the time for a trip when given the average speed and the total distance traveled;

(b) the total distance traveled when given the time for a trip and the average speed; and

(c) the average speed for a trip, no matter what units are used to describe the total distance and the total time of the trip.

Invitation to Inquiry

Have you ever seen an entire stage covered with dominoes lined up, one after another and winding around into interesting patterns? The entertainer tips over one domino, which falls into another, which falls into the one next to it... and on until in a short time all the dominoes have fallen over.

How far apart should the dominoes be spaced for maximum speed? Is it possible to vary this speed by changing the spacing? One domino causes a falling row to continue falling by hitting its neighbor, so the limit to how far apart the dominoes are spaced must be the length of a domino. The other limit would be zero space between two adjacent dominoes, so the limits to the spacing between two adjacent dominoes must be somewhere between zero and one domino length. Thus it would be convenient to record spaces between dominoes as a *ratio* of domino lengths, that is, the space between dominoes in cm divided by the length of one domino in cm.

If you accept this invitation, you will need to determine how you plan to space the dominoes as well as how many dominoes are needed to measure the speed. By making a graph and doing some calculations, can you predict how many dominoes would be needed—and at what spacing—to make a row that takes exactly 2 minutes to fall?

Ratio of spacing length to domino length (spacing/domino length).

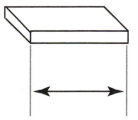

Domino Length
(example 4.0 cm)

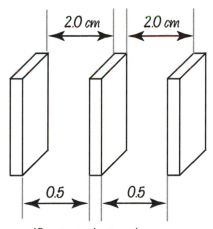

(Spacing in domino ratios:
from 2.0 cm/4.0 cm = 0.5)

Data Table 3.1	Distance and Time Data for Battery-Powered Toy over a Flat Surface
Total Time (s)	Total Distance (cm)
_____	_____
_____	_____
_____	_____
_____	_____
_____	_____
_____	_____

Time (s)	Total Distance (cm)		
	10 cm Elevation	20 cm Elevation	30 cm Elevation
_____	_____	_____	_____
_____	_____	_____	_____
_____	_____	_____	_____
_____	_____	_____	_____
_____	_____	_____	_____

Data Table 3.2   Distance and Time Data for Battery-Powered Toy over Elevated Surfaces

Data Table 3.3	Time and Distance Data for Rolling Ball on Ramp			
Distance from Bottom (cm)	Time Trial 1 (s)	Time Trial 2 (s)	Time Trial 3 (s)	Time Average (s)
_____	_____	_____	_____	_____
_____	_____	_____	_____	_____
_____	_____	_____	_____	_____
_____	_____	_____	_____	_____
_____	_____	_____	_____	_____
_____	_____	_____	_____	_____

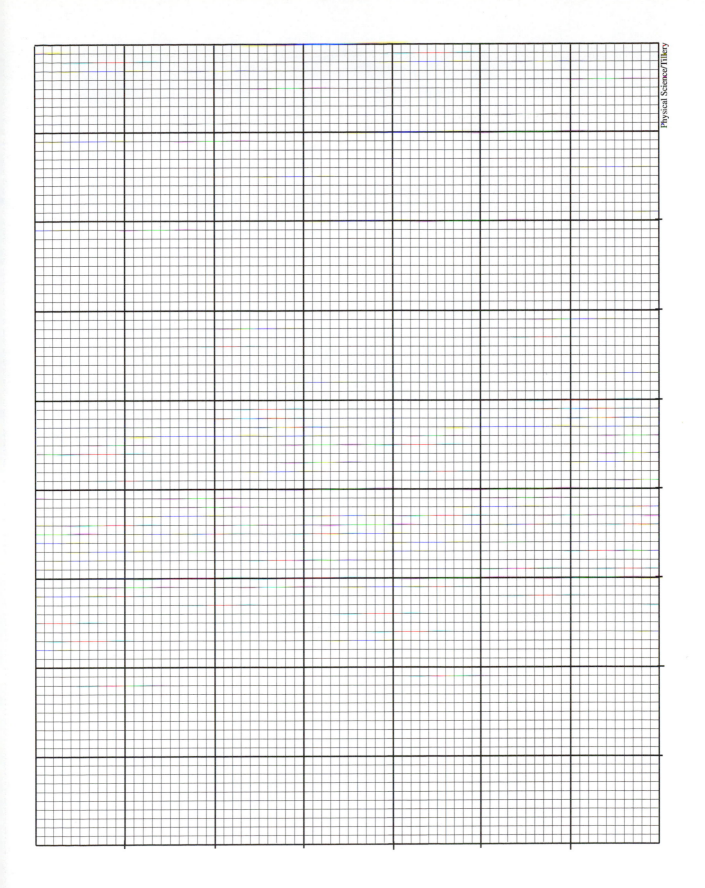

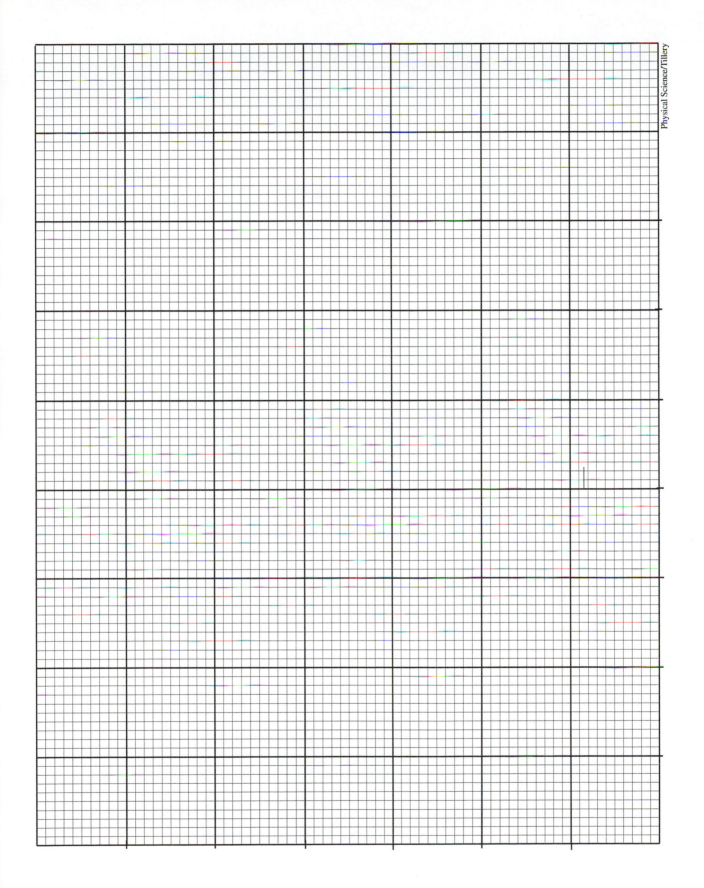

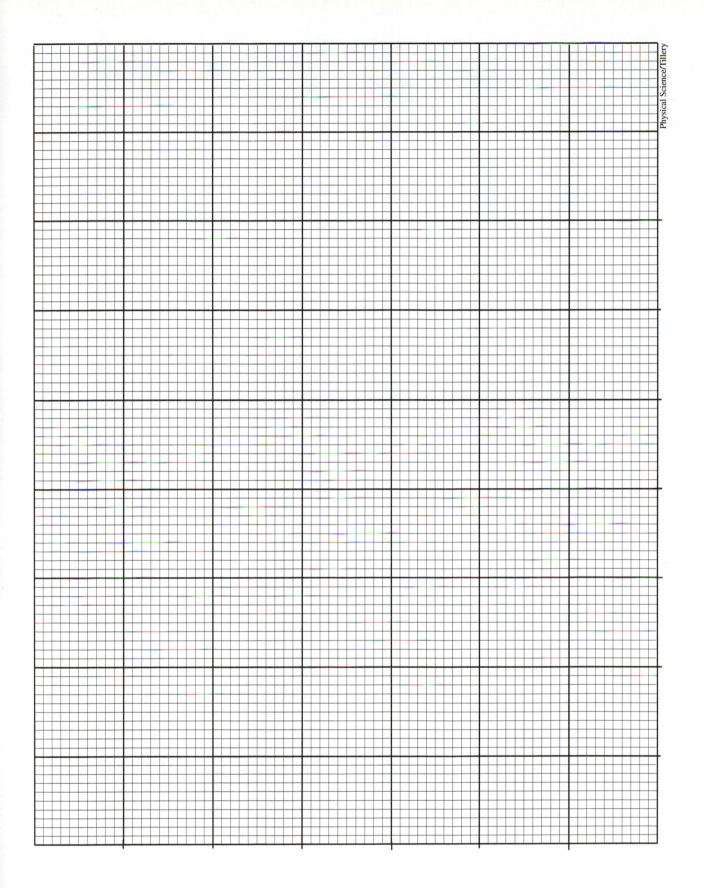

Experiment 4: Free Fall

Introduction

In this experiment you will calculate the acceleration of an object as it falls toward the earth's surface. Air resistance is not considered for an object in *free fall*, and such an object moves toward the surface with a uniform accelerated motion due to gravity, g. The value of g varies with location on the surface of the earth, increasing with latitude to a maximum at the poles. The value of g also varies with elevation, decreasing with elevation at a certain latitude. The average, or standard, value of g, however, is usually accepted as 9.8 m/s^2 or 980 cm/s^2.

When you measure the total distance that an object moves during some period of time, you can calculate an average velocity. **Average velocity** is defined as

$$\overline{v} = \frac{\Delta d}{\Delta t}$$

where Δd is the total distance (final distance minus initial, or $d_f - d_i$) and Δt is the total time (final time minus initial, or $t_f - t_i$). In this experiment you will be measuring the velocity of an object that falls from an initial distance and time of zero, so $\Delta d = d_f - 0$ and $\Delta t = t_f - 0$. For the case of a falling object,

$$\overline{v} = \frac{\Delta d}{\Delta t} = \frac{d_f - d_i}{t_f - t_i} \quad \text{since } d_i = 0 \text{ and } t_i = 0 \quad \therefore \quad \overline{v} = \frac{d_f}{t_f}$$

Thus you can calculate the average velocity of an object in free fall from the total distance traveled and the time of fall. When an object moves with a constant acceleration, you can also find the average velocity by adding the initial and final velocity and dividing by 2,

$$\overline{v} = \frac{v_f + v_i}{2}$$

By substituting the other expression for average velocity, we have

$$\overline{v} = \frac{v_f + v_i}{2} \quad \text{and} \quad \overline{v} = \frac{d_f}{t_f} \quad \therefore \quad \frac{v_f + v_i}{2} = \frac{d_f}{t_f}$$

Since the initial velocity of a dropped object is zero, then v_i is zero and we can solve for the final velocity of v_f, and

$$\frac{v_f}{2} = \frac{d_f}{t_f} \quad \therefore \quad v_f = \frac{2d_f}{t_f}$$

In this experiment you will measure the distance a mass has fallen during recurring time intervals according to a timing device. This data will enable you to calculate the instantaneous velocity at known time intervals. Plotting the velocity versus the time, then finding the slope will provide an experimental value of g.

Procedure

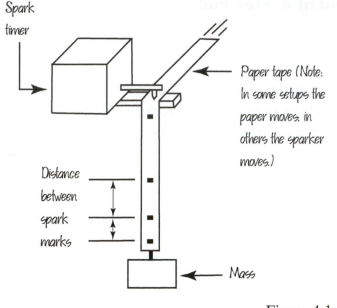

Spark timer

Paper tape (Note: In some setups the paper moves; in others the sparker moves.)

Distance between spark marks

Mass

Figure 4.1

1. You will experimentally determine the acceleration due to gravity and compare it to the standard value of 980 cm/s^2. The procedure may vary with the apparatus used. For example, you might use an apparatus that consists of a device to measure the free-fall of an object with a spark timer that will mark a paper tape at equal time intervals. As a mass accelerates downward it will leave a trail of spark marks at equal time intervals. You will draw a perpendicular line through each mark, then identify the first mark as your reference line. The first mark is identified as the place where $d_f = 0$. Other means of measuring velocity that might be used in your laboratory, such as the use of photogates and computer software, will be explained by your instructor.

2. For spark mark (or ink dot) trails measure the *total distance* (d_f) by using the beginning mark as a reference line. On page 48, record in Data Table 4.1 the distance in cm of each mark *from the reference line*.

3. For each spark, record the *time* (t) that elapsed between the marks as determined by the spark timer. Your instructor will provide exact information for your timer. Most timers are set to operate on 60 Hz, making a spark every 1/60 second. Thus the second spark would have occurred 1/60 second after the first, the third spark mark would have occurred 1/60 plus 1/60, or 2/60 (0.033 s) after the first mark. Fill in Data Table 4.1 with the total distance and time data for each mark, and calculate and record the velocity at each spark. Repeat the experiment two more times with two more paper tapes, completing Data Table 4.2 (page 49) and Data Table 4.3 (page 50).

Results

1. Look over the data in Data Table 4.1, 4.2, and 4.3, think about what the data means, then select the Data Table that seems to have the "best run" data. State which table was chosen and explain the basis for your choice.

2. Using the data table from the best run, make a graph with *velocity* (*v*) on the y-axis and *elapsed time* (*t*) on the x-axis. (Note: Because the first spark was probably not made at the actual time of release, the line on your graph will probably not have a *y* intercept of 0.) Find the slope and record it here, along with any notes you may wish to record.

3. Use the calculated slope and the accepted value of 980 cm/s^2 to calculate the experimental error.

4. Was the purpose of this lab accomplished? Why or why not? (Your answer to this question should show thoughtful analysis and careful, thorough thinking.)

Going Further

What is your reaction time? One way to measure your reaction time is to have another person hold a meterstick vertically from the top while you position your thumb and index finger at the 50 cm mark. The other person will drop the meterstick (unannounced) and you will catch it with your thumb and finger. Accelerated by gravity (g), the stick will fall a distance (d) during your reaction time (t). Knowing d and g, all you need is a relationship between g, d, and t to find the time.

You know a relationship between d, $\bar{v}$, and t from $\bar{v} = d / t$. Solving for d gives $d = \bar{v}t$.

Any object in free-fall, including a meterstick, will have uniformly accelerated motion, so the average velocity is

$$\bar{v} = \frac{v_f + v_i}{2}$$

Substituting for the average velocity in the previous equation gives

$$d = \left(\frac{v_f + v_i}{2}\right)(t)$$

The initial velocity of a falling object is always zero just as it is dropped, so the initial velocity can be eliminated, giving

$$d = \left(\frac{v_f}{2}\right)(t)$$

Now you want acceleration in place of velocity. From

$$a = \frac{v_f - v_i}{t}$$

and solving for the final velocity gives

$$v_f = at$$

The initial velocity is again dropped since it equals zero. Substituting the final velocity in the previous equation gives

$$d = \left(\frac{at}{2}\right)(t) \qquad or \qquad d = \frac{1}{2}at^2$$

Finally, solving for t gives

$$t = \sqrt{\frac{2d}{g}}$$

Measuring how far the meterstick falls (in m) can now be used as the distance (d) with g equaling 9.8 m/s^2 to calculate your reaction time (t).

Invitation to Inquiry

Find out how well you can predict the motion of falling objects. First, select some objects such as a rubber ball, a sheet of notebook paper, and a large metal paper clip. Predict, then study the detail of each object falling independently... for example, what happens to each as they fall? Then compare the motion of the objects side by side. Is it possible to cause them to fall together, at the same time?

Here are the five basic shapes of graphs that are produced by different types of relationships between variables. What variables are involved in falling objects? What are the relationships between them? If any of the graphs you produce have the shapes of one or more of the curves, what can be done to make it into a straight line?

Use measurements to construct a graph or graphs that show what is going on between the variables involved in falling objects. Then use the graph to show how to place three or four objects on a long cord. Attach them so when the cord is hung from a high place, then dropped, the objects make a constant plop, plop, plop sound when they hit the ground.

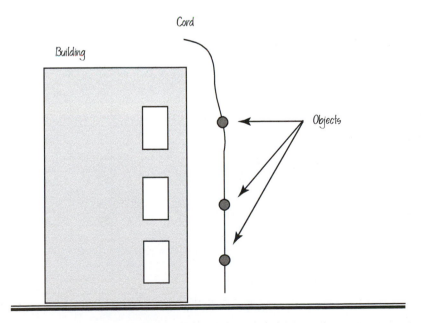

47

Data Table 4.1   Free Fall Run Number One

Spark Number	Distance (cm)	Time of Fall (s)	Computed Instantaneous Velocity* (cm/s)
1			
2			
3			
4			
5			
6			
7			
8			
9			
10			

*From $v_f = \dfrac{2d_f}{t_f}$

Data Table 4.2	Free Fall Run Number Two		Computed Instantaneous Velocity* (cm/s)
Spark Number	Distance (cm)	Time of Fall (s)	
1			
2			
3			
4			
5			
6			
7			
8			
9			
10			

*From $v_f = \dfrac{2d_f}{t_f}$

Data Table 4.3   Free Fall Run Number Three

Spark Number	Distance (cm)	Time of Fall (s)	Computed Instantaneous Velocity* (cm/s)
1			
2			
3			
4			
5			
6			
7			
8			
9			
10			

*From $\quad v_f = \dfrac{2d_f}{t_f}$

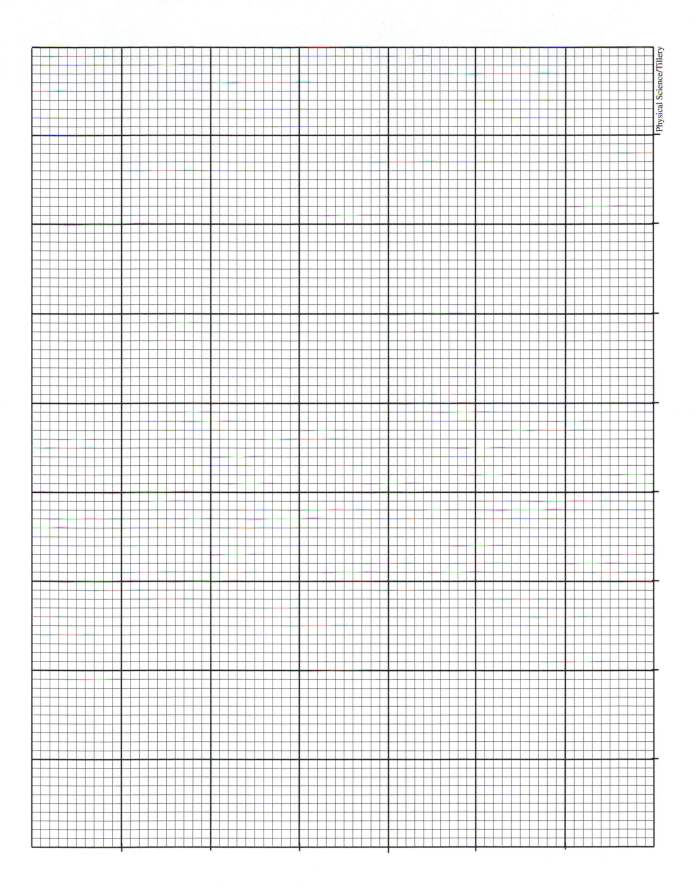

Experiment 5: The Pendulum

Introduction

A simple pendulum consists of a mass, called a **bob**, connected to the end of a suspended cord or string. When the bob is pulled to one side of its position of rest and then released, it begins to vibrate in a simple harmonic motion. In general, a **vibration** is any back and forth motion that repeats itself. Simple harmonic motion occurs when there is a restoring force that is equal and opposite to a displacement. The restoring force, of course, comes from gravity acting on the bob.

There are a number of factors that could affect the vibration of a pendulum and this investigation is concerned with these factors. A vibrating pendulum can be described by measuring several variables. The extent of displacement from the rest position is called the **amplitude** of the vibration (C to A – or B – in figure 5.1). A vibration that has the bob displaced a greater distance from the rest position thus has a greater amplitude than a vibration with less displacement.

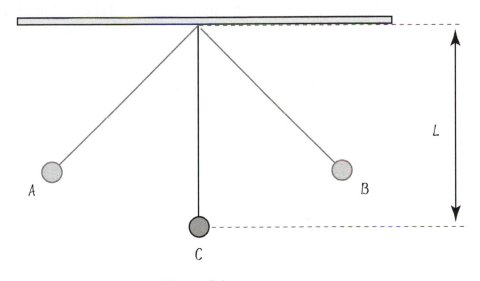

Figure 5.1

A complete vibration is called a cycle. A **cycle** is the movement from some point (say the far left), to a maximum displacement in the other direction (say the far right) then back to the same point again (the far left). The **period** (T) is simply the time required to complete one cycle. For example, suppose 3 s is required for a bob to move through one complete cycle, to complete the back-and-forth motion from one point, then back to that point. The period of this vibration is 3 s. The number of cycles per second is called the **frequency** (f). For example, a vibrating bob that moves through 1 cycle in 1 s has a frequency of 1 cycle per second. The **length** (L) of the pendulum is measured from the point of suspension to the center of gravity of the bob. Ideally the mass of the bob should be concentrated at a single point. For a small homogeneous sphere, however, the distributed mass is

53

affected by gravity almost as if the mass were all located at the center of the sphere. For all practical purposes then, the length of the pendulum is the *length of the cord plus the radius of the bob* (see figure 5.1).

Procedure

1. Investigate if *weight* influences the period of a pendulum. Use different masses for bobs and adjust each so the length of the string is 100 cm from the pivot point to the center of gravity of the bob. Pull the bob back to make an arc of about 15 degrees. Release and count the number of vibrations (one vibration is one complete back and forth movement) for exactly one minute. The arc, length of pendulum, and all other variables must be the same for each run; the only difference should be the mass of the bob. Take three trials for each mass and average the findings. Record all data in Data Table 5.1 on page 59.

2. This time, hold all the variables the same except the *amplitude* of the bob. You can measure the approximate amplitude by using two meter sticks as shown in figure 5.2. Count the number of vibrations for exactly one minute for each amplitude tested. Take three trials for each amplitude, average the findings, and record all data in Data Table 5.2 on page 60.

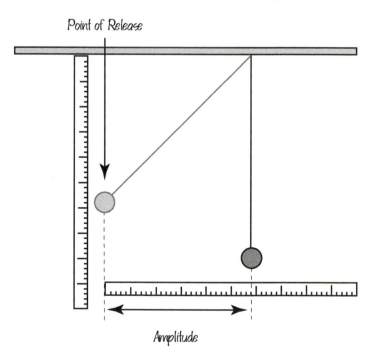

Figure 5.2

3. This time hold all the variables the same except the *length* of the pendulum. Recall that the length of the pendulum is not just the length of the cord. The length is measured from the pivot point of the pendulum to the center of gravity of the bob. Measure the number of vibrations for exactly one

54

minute. Repeat for three trials for each length, average the findings, and record all data in Data Table 5.3. From the average data, calculate the experimental period for each pendulum length (the time required for one complete vibration) and record it in Data Table 5.3. Use the following equation to calculate the theoretical value of the period for each length and record it on page 61 in Data Table 5.3:

$$T = 2\pi\sqrt{\frac{L}{g}}$$

4. Use a metal bob to make a pendulum with a 25 cm length. Make three separate measurements of the period of this pendulum by finding the time for the pendulum to vibrate 25 times. Record this data in seconds to the nearest hundredth in Data Table 5.4 on page 62. Multiply the time by 4 to obtain the calculated time of 100 full swings, recording the data in the table. The period in seconds can now be calculated by moving the decimal two places to the left. Use this period and the length of the pendulum in the following equation to calculate the acceleration of the pendulum due to gravity. Then calculate the percentage error and record it in the table.

$$g = \frac{4\pi^2 L}{T^2}$$

Results

1. According to your experimental results, what effect does the weight have on the period of a pendulum?

2. According to your experimental results, what effect does the amplitude have on the period of a pendulum?

3. According to your experimental results, what effect does the length have on the period of a pendulum?

4. Offer a theoretical explanation for all three of the results described for questions 1-3.

5. Was the purpose of this lab accomplished? Why or why not? (Your answer to this question should be reasonable and make sense, showing thoughtful analysis and careful, thorough thinking.)

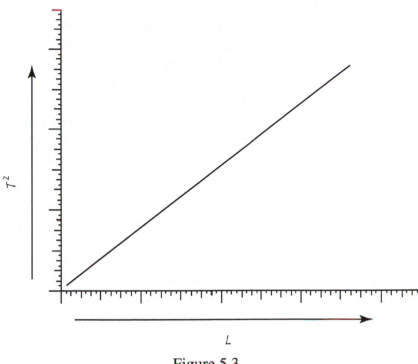

Figure 5.3

If you were to plot T^2 and L from Data Table 5.3, you would see the relationship illustrated in figure 5.3. The straight-line relationship represented in figure 5.3 can be written as

$$T^2 = kL$$

where k is the slope of the line. The true relationship is

$$T = 2\pi\sqrt{\frac{L}{g}}$$

Therefore, k is equal to

$$\frac{4\pi^2}{g}$$

Use the three data points from procedure step 4 (Data Table 5.3) to plot T^2 vs. L, then see if the slope is equal to $4\pi^2/g$. Describe your findings here.

Invitation to Inquiry

What is the acceleration of a pendulum bob as it swings back and forth? One way to find out is to measure acceleration just as Galileo did hundreds of years ago. Galileo found that you could "slow" the acceleration of a falling object by making them "fall" at an angle, that is, fall down an inclined plane. One way to measure the acceleration of a pendulum bob is to set up an inclined plane so a rolling ball "falls" down the board at the same angle as it "falls" as a pendulum.

First, obtain a ramp or board about two meters long. Measure down from the top about 20 cm and draw a zero line. Place tape marks at 25 cm, 50 cm, 100 cm, and 150 cm down from the zero line. Use a pencil to hold the ball at the zero line then pull it away quickly to release the ball. Place a wooden block or some other stop at the bottom to keep the ball from rolling off the ramp. Figure out a procedure for using a stopwatch to calculate the acceleration of the ball at each of the marks while a ball is rolling down the ramp. Should the acceleration change as the ball rolls downhill?

Second, move the ramp so the angle is about the same as a model pendulum you have set up. Adjust the ramp so the ball takes about the same time to roll down the ramp as the pendulum bob does to swing from its high to low point. This comparison will take care of friction losses on the ramp. What is the acceleration of the ball? Can you find a pattern to changes in acceleration and changes in ramp angle? Finally, set up an experiment to compare the findings of the pendulum experiment with the acceleration of various weight balls rolling down an incline plane. What did you expect to find?

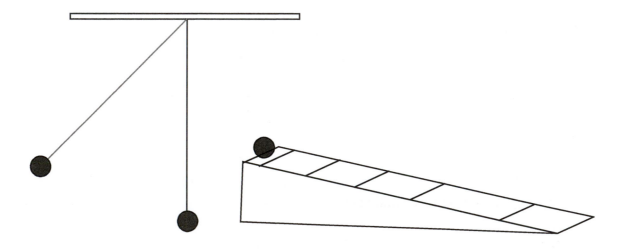

Bob	Cycles per Minute			
	Trial 1	Trial 2	Trial 3	Average
Name: Plastic Mass: (g)	_____	_____	_____	_____
Name: Wood Mass: (g)	_____	_____	_____	_____
Name: Mass: (g)	_____	_____	_____	_____
Name: Mass: (g)	_____	_____	_____	_____
Name: Mass: (g)	_____	_____	_____	_____
Name: Mass: (g)	_____	_____	_____	_____

Data Table 5.1   The Effect of Weight on a Pendulum

Data Table 5.2	The Effect of Amplitude on a Pendulum			
Approximate Amplitude	Cycles per Minute			
	Trial 1	Trial 2	Trial 3	Average
(cm)				
_____	_____	_____	_____	_____
_____	_____	_____	_____	_____
_____	_____	_____	_____	_____
_____	_____	_____	_____	_____
_____	_____	_____	_____	_____
_____	_____	_____	_____	_____

Length of Cord plus Radius of Bob	Cycles per Minute					
	Trial 1	Trial 2	Trial 3	Average	Experimental Period	Calculated Period
(cm)					(s)	(s)
————	————	————	————	————	————	————
————	————	————	————	————	————	————
————	————	————	————	————	————	————
————	————	————	————	————	————	————
————	————	————	————	————	————	————
————	————	————	————	————	————	————

Data Table 5.3   The Effect of Length on a Pendulum

Data Table 5.4   The Pendulum and Acceleration Due to Gravity

Trial	Time of 25 full swings (s)	Calculated Time of 100 full swings (s)	Period (T) (s)	Square of Period (T²) (s)
Example	25.10	100.4	1.004	1.01
1				
2				
3				
Average				

Experimental Value of g: _____

Theoretical Value of g: 980 cm/s²

Percentage Error:

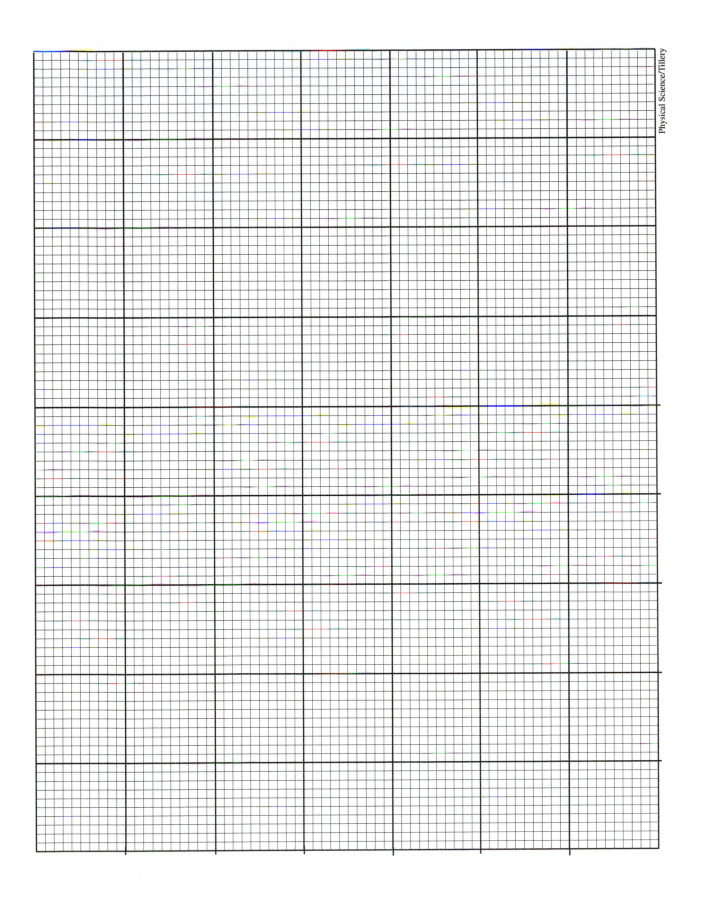

Experiment 6: Projectile Motion

Introduction

There are basically three kinds of motion: (1) the *horizontal*, straight-line motion of objects moving on the surface of the earth; (2) the *vertical* motion of dropped objects that accelerate toward the surface of the earth; and (3) the motion of an object that is projected into the air. The third type of motion, **projectile motion**, could be directly upward as a vertical projection, straight out as a horizontal projection, or at some angle between the vertical and the horizontal. Basic to understanding such compound motion is to understand that (1) gravity always acts on objects, no matter where they are, and (2) the acceleration due to gravity (g) is independent of any motion that an object may have.

As an example of projectile motion, consider figure 6.1. After rolling down the incline AB, the ball moves across a frictionless, horizontal track BC. At C the ball leaves the track to become a projectile. While the ball is still on track BC, and ignoring air resistance, the *speed* of the ball on the track is *constant* because there are *no net forces acting on the ball*. ($F = ma$; if $F = 0$, then $a = 0$.)

After the ball leaves the track, it becomes a projectile. The motion of such a projectile is easier to understand if you split the complete motion into vertical and horizontal parts. After the ball leaves the track, there is an unbalanced force (mg) that accelerates the ball vertically downward. The ball thus has an *increasing downward velocity* the same as that of a dropped ball and is represented by the vertical vector arrows (v_y) in Figure 6.2. Ignoring air resistance, there are no forces in the

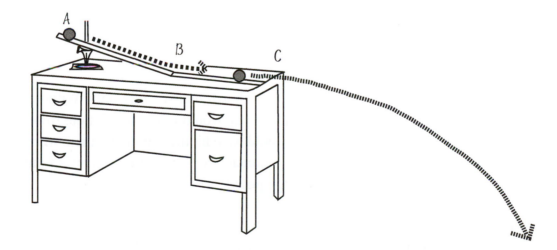

Figure 6.1

horizontal direction so the *horizontal velocity remains the same* as shown by the v_x arrows. The combination of the vertical motion (v_y) and the horizontal motion (v_x) causes the ball to follow a curved path until it hits the floor.

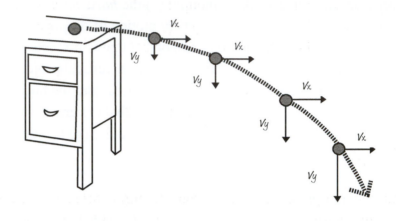

Figure 6.2

The *vertical distance* (Δy) that a falling object moves is proportional to the square of the time that it is falling vertically (t_y). Considering the acceleration due to gravity (g), then

$$\Delta y = \frac{1}{2}gt_y^2$$

The *horizontal distance* (Δx) that a ball moves depends on its horizontal velocity (v_x) when it leaves the horizontal track. Velocity is distance per unit time ($v = d/t$), so

$$v_x = \Delta x/t_x \text{ and } t_x = \Delta x/v_x$$

Imagine what would happen if the ball had a horizontal velocity only. Without gravity there would be no increasing downward velocity and the ball would move straight out from the table. It would, however, be vertically above where it would have hit the ground (if there were a downward velocity) at the same time. You can see this if you mentally remove the v_y arrows from figure 6.2. Thus the time of fall is determined by t_y, and t_x equals t_y, or $t_x = t_y$.

Since $t_x = t_y$ and $t_x = \Delta x/v_x$, you can substitute $\Delta x/v_x$ for t_y in $\Delta y = \frac{1}{2}gt_y^2$. Then the expression is $\Delta y = \frac{1}{2}g(\Delta x/v_x)^2$ or simply $\Delta y = \frac{1}{2}g(\Delta x^2/v_x^2)$, which now has both the vertical distance (Δy) and the horizontal distance (Δx) in the same relationship. Look at figure 6.3 and think about this relationship.

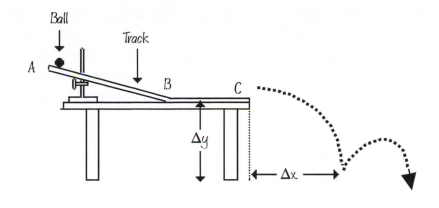

Figure 6.3

Procedure

1. Adjust the ramp so that A, the uppermost position of the ramp, is 10 cm above the surface of the table top. Measure the horizontal part of the ramp labeled BC. Release the ball from A and time how long the ball takes to move across the distance BC. Find the velocity (v_x) of the ball from the relationship $v = d/t$. Make at least three runs and find the average velocity. Record your data in Data Table 6.1.

2. Measure and record Δy. Solve $\Delta y = \frac{1}{2}g(\Delta x^2/v_x^2)$ for Δx, then use the values of Δy and v_x from Data Table 6.1 to find Δx. Record what Δx should be according to your calculation.

3. Place a cup on the floor at the calculated distance Δx from the edge of the table. Roll the ball down the ramp to see if your calculated prediction was correct. You can also use a piece of carbon paper on a sheet of paper for a target.

4. Repeat procedure steps 1 through 3 for ramp heights of 20 cm and 30 cm, making sure that the position of the stand from the edge of the table is constant.

Results

1. Make a graph with velocity on the *x*-axis and distance on the *y*-axis, and use the average v_x velocity and the average Δx distance from each data table as data points. Calculate the slope, including units, and write the value of the slope here and also somewhere on the graph.

2. Calculate t_y from the square root of $2\Delta y/g$, and discuss why this value should or should not be equal to the slope obtained in question 1 above.

3. Use the results from your calculation in question 2 as the accepted value, and the value of the slope calculated in question 1 as the experimental value, to determine your percentage error. What was the percentage error and probable sources?

4. Discuss how you could improve the precision of this experiment.

5. Was the purpose of this lab accomplished? Why or why not? (Your answer to this question should show thoughtful analysis and careful, thorough thinking.)

Invitation to Inquiry

In this investigation you studied a relationship between the velocity of a moving ball and the distance it travels when rolled off a table. Why not consider the height from which the ball is released and the distance it travels before it hits the floor? First, move the ramp so the end of the track is parallel to the table top and the ball shoots off the track horizontally. This might require a curved ramp, depending on your setup. Move the apparatus so that the end of the track is flush with the edge of the table (see diagram below). Again use a sheet of carbon paper on a sheet of white paper so the ball will leave a mark you can measure. Release the ball from at least 6 different heights, where height is the measure of how far you raise the ball above the flat portion of the track. For each height measure the horizontal distance which the ball traveled. Graph your results, then do what you need in order to straighten out at a curve if one is obtained. What do your findings mean about the distance and height relationship; what is the relationship?

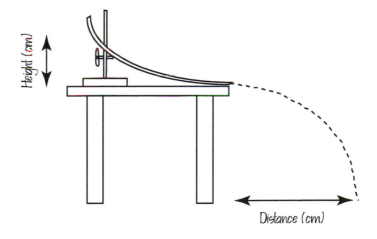

Data Table 6.1	Projectile Motion from a Ramp: Ramp 10 cm Above Table Top	
Trial	Time (t) (s)	Velocity (v) (cm/s)
1	——	——
2	——	——
3	——	——

Average v_x _____

Vertical distance Δy _____

Calculated Δx _____

Trial	Time (t) (s)	Velocity (v) (cm/s)
Data Table 6.2 Projectile Motion from a Ramp: Ramp 20 cm Above Table Top		
1	_____	_____
2	_____	_____
3	_____	_____

Average v_x _____

Vertical distance Δy _____

Calculated Δx _____

Data Table 6.3	Projectile Motion from a Ramp: Ramp 30 cm Above Table Top	
Trial	Time (t) (s)	Velocity (v) (cm/s)
1	_____	_____
2	_____	_____
3	_____	_____

Average v_x _____

Vertical distance Δy _____

Calculated Δx _____

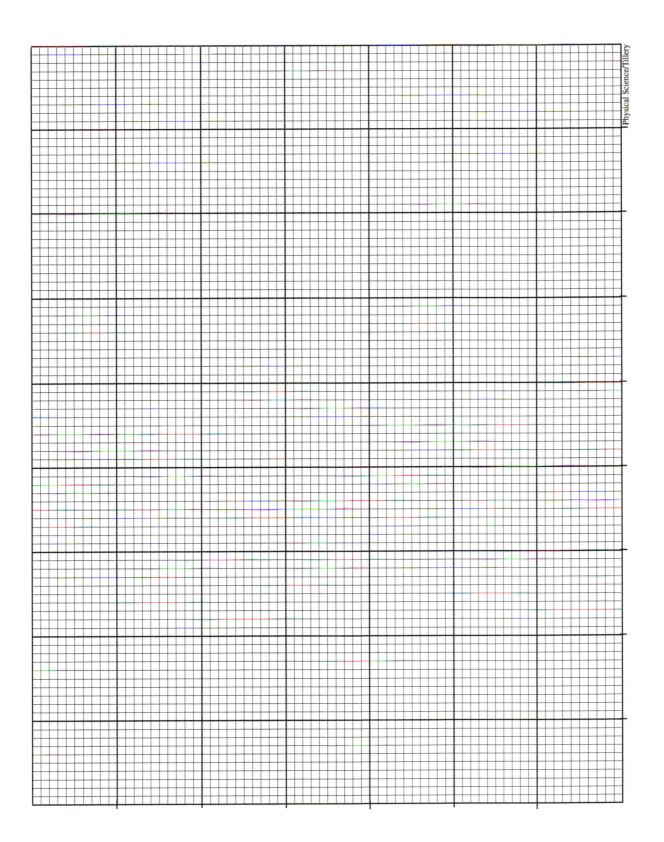

Experiment 7: Newton's Second Law

Introduction

In this laboratory experiment you will consider **Newton's second law of motion**, which states that an object will accelerate if an unbalanced force acts on it. Upon successful completion of this experiment you should be able to

1. determine the acceleration of a system with a constant mass and a varying force;

2. calculate the mass of the system from experimental data by knowing that the slope of the graph is 1/mass; and

3. calculate the percentage error between the experimental result and the accepted value for a mass.

Newton's second law of motion states the relationship that can be used to predict the change of motion (acceleration) of an object when an unbalanced force is applied. The acceleration is directly proportional to the force and inversely proportional to the mass of the object, or $a = F/m$.
Recall that the slope of a straight line is a constant obtained from

$$\text{slope } (m) = \frac{\Delta y}{\Delta x}$$

This equation can be solved for y, with m representing the slope of the straight line. When the origin is zero, the solved equation is as shown in figure 7.1.

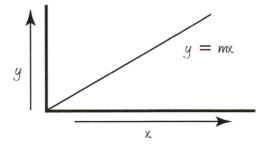

Figure 7.1

From this, you can see that y is directly proportional to x; that is, any increase in x (the manipulated variable) will result in a proportional increase in y (the responding variable).
Figure 7.2 shows three different relationships between three different sets of x and y variables. Each slope has a different proportionality constant (m_1, m_2, and m_3) and each relationship is represented by the equation $y = mx$. In each case the proportionality constant in the equation is equal to the slope of the line.

75

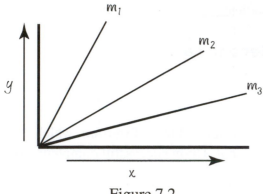

Figure 7.2

In this experiment, you will determine the acceleration of an object by varying the force and keeping the mass constant. From Newton's second law, $F = ma$, you know that

$$\text{acceleration} = \frac{\text{Force}}{\text{mass}} \quad \text{or} \quad a = \frac{F}{m} \quad \text{or} \quad a = \left(\frac{1}{m}\right)(F)$$

This equation is in the same form as the equation for a straight line (when the origin is equal to zero), or

$$y = mx$$

and

$$a = \left(\frac{1}{m}\right)(F)$$

where y = acceleration, x = force, and the slope (m) is equal to $1/m$.

Again, recall that the slope of a straight line is determined from

$$\text{slope } (m) = \frac{\Delta y}{\Delta x}$$

If you plot acceleration (m/s^2) on the y-axis and force (N) on the x-axis, the units for the slope would be

$$\frac{y = \text{m}/\text{s}^2}{x = \text{newtons}} =$$

$$\frac{\text{m}/\text{s}^2}{\text{kg}\cdot\text{m}/\text{s}^2} = \frac{\text{m}}{\text{s}^2} \times \frac{\text{s}^2}{\text{kg}\cdot\text{m}} = \frac{1}{\text{kg}}$$

The slope therefore is the *reciprocal* of the mass, or 1/kg. As shown in figure 7.3, this is just what you would expect for

$$a = \left(\frac{1}{m}\right)(F)$$

76

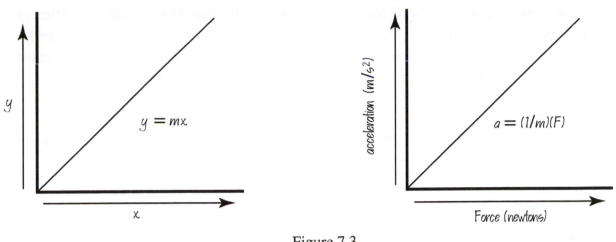

Figure 7.3

Procedure

1. The method of measuring cart or glider velocity may vary with the laboratory setup. You might use a system of a horizontally moving cart attached to weights on a hanger through a string over a pulley. As the hanger falls, it exerts a horizontal force of $F = mg$ on the cart. Varying the mass on the hanger will thus vary the horizontal force on the cart. Other equipment with other means of measuring velocity might be used in your laboratory, such as a cart with wheels and the use of photogates and computer software. If so, the use of the apparatus will be explained by your instructor.

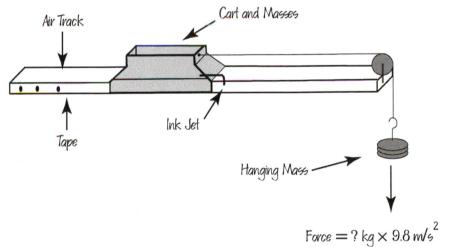

Figure 7.4

2. Tie a nylon string onto the cart. The string should hang about 30 cm down from the pulley when the cart is in the starting position. A weight hanger is attached to the hanger string.

3. Arrange five 50.0 g masses on the cart so they are evenly distributed. You will make six separate runs. On the first run the force on the cart will be the mass of the hanger only times g. For each consecutive run a 50 g mass *is removed from the cart and placed on the hanger*. This

keeps the mass of the system constant but varies the force on the cart. (Note: There are different ways to define a system, and for this lab the system is defined as the cart, masses, and hanger). According to the definitions used, the mass of the system is kept constant (while varying the force) by moving masses from the cart and placing them on the hanger. You could add masses to the hanger from outside the system defined, but that is a different experiment.

4. For each of the six runs, calculate the force on the cart from the mass of the hanging masses and hanger in kg times 9.8 m/s^2. The force will therefore be calculated in *newtons*. Record the force for each run in the appropriate data table.

5. Measure the position of the dots on the tape for each of the six runs. The timer will be set so the dots are 0.05 seconds apart. Carefully measure the distances between the dots and use the change in distance and the time elapsed to calculate the velocity, change in velocity, and acceleration for each run. Record these quantities in the appropriate data tables.

6. Find the average value of the acceleration (a) for each run. Note that the average value must be converted to m/s^2 so the units will be consistent with the units of a newton.

7. Record the average acceleration and the calculated force for each run in the summary data table on page 84. Record the mass of the system (cart, hanger, and masses) in the summary data table.

8. Use the average acceleration and the force values in the summary data table for data points on a graph, with the acceleration in m/s^2 on the *y*-axis and the force in newtons on the *x*-axis. You will have six data points, with each data point corresponding to a run. Calculate the slope of the line (numbers and units) and write it somewhere on the graph.

Results

1. Using the reciprocal of your calculated slope as the experimental value of the mass of the system and the measured mass of the system (cart, hanger, and masses) as found by a balance as the accepted value, calculate the experimental error. Show your calculations here.

2. How do you know if the acceleration was proportional to the force? (Hint: see figure 7.3)

3. How would you know if the acceleration was inversely proportional to the mass?

4. Would the presence of an additional mass in the cart during all the runs increase, decrease, or not affect the slope? How could you use this experimental set-up to find the value of the additional mass that was in the cart?

5. Suppose the string breaks as the cart accelerates one-third of the way across the track. What is the acceleration of the cart for the remaining length of track? Explain.

6. Was the purpose of this lab accomplished? Why or why not? (Your answer to this question should be reasonable and make sense, showing thoughtful analysis and careful, thorough thinking.)

Invitation to Inquiry

1. Design a demonstration of Newton's second law of motion using a student on ice skates (or roller blades) and a spring scale. The student on skates or roller blades should hold the hook on one end of the scale while you pull the hook on the other end, pulling with a sufficient force to exactly maintain a constant force. Calculate the acceleration by using the equation $F = ma$, then measure the acceleration a second way by marking distances and measuring the change of velocity.

2. Repeat the demonstration with a more massive student on ice skates or roller blades. Compare the difference, if any, needed to maintain a constant force on a more massive moving object.

3. Make a velocity vs. time graph for the application of a constant force on the first student, and another graph for the more massive student. What characteristic of the graph changes?

4. Using any everyday materials you wish, design a demonstration to *prove* that a moving object with zero net force moves at a constant velocity.

5. For the demonstration in Invitation 1 (constant force) and the demonstration in Invitation 4 (zero force) sketch three graphs to predict (a) force vs. time; (b) acceleration vs. time; and, (c) force vs. acceleration. Do several demonstrations to show your six predictions are correct.

Data Table 7.1   Run 1

	d (cm)	Δd (cm)	Δt (s)	$v = \Delta d / \Delta t$ (cm/s)	Δv (cm/s)	$a = \Delta v / \Delta t$ (cm/s^2)
1						
2						
3						
4						
5						

Force = _____(kg)(9.8 m/s^2) = _____N

Average acceleration = _____(total)/_____(number) = _____cm/s^2

Average acceleration = _____ m/s^2

Data Table 7.2   Run 2

	d (cm)	Δd (cm)	Δt (s)	$v = \Delta d / \Delta t$ (cm/s)	Δv (cm/s)	$a = \Delta v / \Delta t$ (cm/s^2)
1						
2						
3						
4						
5						

Force = _____(kg)(9.8 m/s^2) = _____N

Average acceleration = _____(total)/_____(number) = _____cm/s^2

Average acceleration = _____ m/s^2

Data Table 7.3   Run 3

	d (cm)	Δd (cm)	Δt (s)	$v = \Delta d/\Delta t$ (cm/s)	Δv (cm/s)	$a = \Delta v/\Delta t$ (cm/s^2)
1						
2						
3						
4						
5						

Force = _____(kg)(9.8 m/s^2) = _____N

Average acceleration = _____(total)/_____(number) = _____cm/s^2

Average acceleration = _____ m/s^2

Data Table 7.4   Run 4

	d (cm)	Δd (cm)	Δt (s)	$v = \Delta d/\Delta t$ (cm/s)	Δv (cm/s)	$a = \Delta v/\Delta t$ (cm/s^2)
1						
2						
3						
4						
5						

Force = _____(kg)(9.8 m/s^2) = _____N

Average acceleration = _____(total)/_____(number) = _____cm/s^2

Average acceleration = _____ m/s^2

Data Table 7.5		Run 5					
		d (cm)	Δd (cm)	Δt (s)	$v = \Delta d/\Delta t$ (cm/s)	Δv (cm/s)	$a = \Delta v/\Delta t$ (cm/s^2)
1							
2							
3							
4							
5							

Force = _____(kg)(9.8 m/s^2) = _____N

Average acceleration = _____(total)/_____(number) = _____cm/s^2

Average acceleration = _____ m/s^2

Data Table 7.6		Run 6					
		d (cm)	Δd (cm)	Δt (s)	$v = \Delta d/\Delta t$ (cm/s)	Δv (cm/s)	$a = \Delta v/\Delta t$ (cm/s^2)
1							
2							
3							
4							
5							

Force = _____(kg)(9.8 m/s^2) = _____N

Average acceleration = _____(total)/_____(number) = _____cm/s^2

Average acceleration = _____ m/s^2

Data Table 7.7	Summary		
Run	Average Acceleration (m/s^2)	Force (N)	Mass of System (kg)
1			
2			
3			
4			
5			
6			

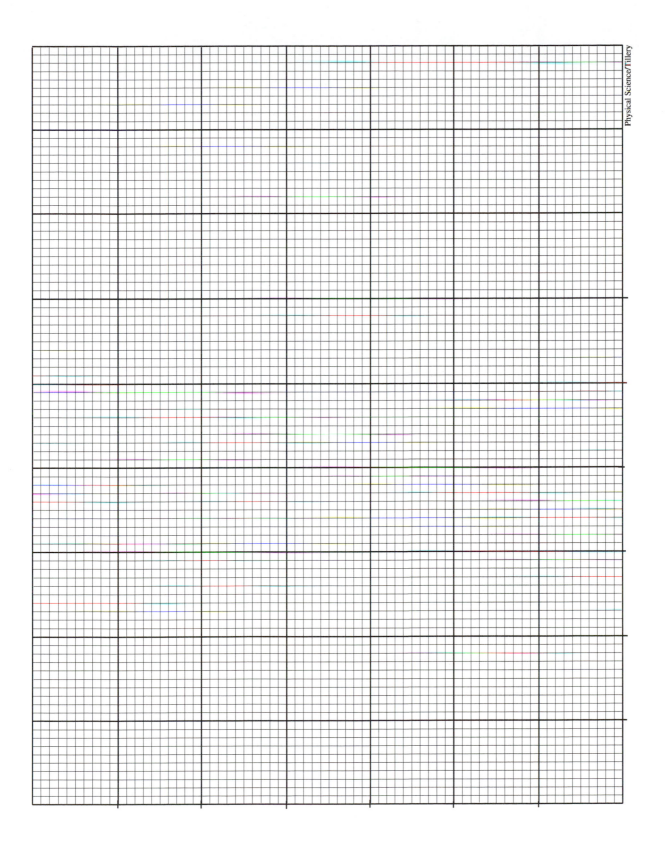

Experiment 8: Conservation of Momentum

Introduction

Linear momentum is defined as the product of the mass of an object and its velocity, or

$$p = mv$$

where p is the symbol for momentum, m is the mass of an object, and v is its velocity. As you can see from this definition, both mass and velocity contribute to the momentum of an object.

Momentum is proportional to the mass and the velocity of an object, which means something must be moving to have momentum (zero velocity times the mass equals zero). If you assume a constant mass of a given object, this means a momentum change of the object is related to a change of velocity. Since a change of velocity (Δv) during a time interval (t) is the definition of acceleration, you could write Newton's second law of motion ($F = ma$) as

$$F = m \frac{\Delta v}{t}$$

Multiplying both sides of the equation by t gives

$$Ft = m\Delta v$$

Since momentum is defined as the product of the mass and velocity ($p = mv$), this means that a change of momentum (Δp) is equal to mv. Thus $Ft = \Delta p$, and

$$F = \frac{\Delta p}{t}$$

So a force must be applied for some time interval for a change of momentum to occur. This also means that no net force means that no change of momentum occurs, $F = 0$ means $\Delta p = 0$. It is a force applied over time that is needed to change the momentum of an object.

When objects are viewed as a system of interacting objects, the total momentum before the interaction is always equal to the total momentum after the interaction. This is expressed as the **law of conservation of momentum**, which states the total momentum of a system of interacting objects is conserved as long as no external forces are involved. For example, the firing of a bullet from a rifle and the recoil of the rifle is a change of momentum in opposite directions. Considering the rifle and the bullet as a system, both the rifle and the bullet have a total momentum of zero with respect to the surface of the earth. When the rifle is fired the exploding gunpowder propels the bullet out with what we could call a forward momentum. At the same time the force from the exploding gunpowder pushes the rifle backward with a momentum opposite to the bullet. The bullet moves forward with a momentum of $(mv)_b$ and the rifle moves backward at the same time. The rifle moves in an opposite direction to the bullet, so its momentum is shown with a minus sign, or $-(mv)_r$. The total momentum of the rifle-bullet system is zero, so

$$\text{Bullet momentum} = -\text{rifle momentum}$$
$$(mv)_b = -(mv)_r$$
$$(mv)_b - (mv)_r = 0$$

the negative sign simply means a momentum in a direction opposite to the other, and that the momentum of the bullet $(mv)_b$ must equal the momentum of the rifle $-(mv)_r$ in the opposite direction.

The law of conservation of momentum also applies in collisions, such as a speeding car colliding with a stationary car of equal mass, with the coupled cars moving together after the collision. Since momentum is conserved, the total momentum of the system of cars should be the same before and after the collision. Thus

$$\text{Momentum before} = \text{momentum after}$$
$$\text{car 1} + \text{car 2} = \text{coupled cars}$$
$$mv_1 = (m+m) \times \frac{v_1}{2}$$
$$mv_1 = mv_1$$

(Note that car 2 had zero momentum with a velocity of zero, so there is no mv_2 on the left side of the equation.).

Procedure

In this experiment you will measure the momentum of laboratory carts involved in inelastic collisions. The experiment could be conducted with gliders moving on an air track (figure 8.1), laboratory carts moving on a track cart, or perhaps other types of carts moving across the floor. Your instructor will explain the operation of the equipment you will use.

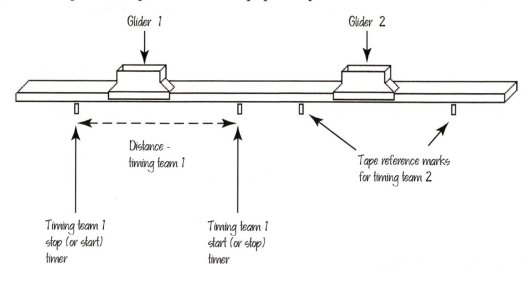

Figure 8.1

The method of measuring cart or glider velocity will vary with the laboratory setup. As shown in figure 8.1, the velocity of a glider or cart before a collision could be measured by measuring the time required to move over a measured distance. Two pieces of tape are used to mark a measured distance *before* the place where a collision will occur. Two other pieces of tape are used to measure the same distance on the track *after* the place where the collision occurs. One team of students, team 1, starts their timers when the front of the cart reaches the first "before collision" reference mark, then stops the timer when the front of the cart reaches the second "before collision" mark. The before collision velocity can be determined from $v = d/\Delta t$. A second team of students, team 2 in figure 8.1, does the same procedure for calculating the "after collision" velocity. Other means of measuring velocity that might be used in your laboratory, such as a spark timer or the use of photogates and computer software will be explained by your instructor.

Part A: Equal Mass Collision

1. Find the mass of a cart (or glider) (m_2) and record the mass in Data Table 8.1 on page 92. Place this cart at rest near the center position of the track.

2. Find the mass of another cart (m_1) and record in Data Table 8.1. Place this nearly equal-sized mass near the end of the track.

3. Conduct a trial run by pushing m_1 **gently but firmly,** giving it an initial velocity toward m_2 at rest. The cart should have its initial velocity *before* reaching the first reference mark tape (if this method is being used). After the collision, m_1 should be at rest and m_2 should be in motion.

4. Conduct three trials of giving m_1 an initial velocity toward m_2 at rest. Record the velocity before and after each collision and record in Data Table 8.1. Calculate the momentum of m_1 before the collision and the momentum of m_2 after each collision. Record these calculations in Data Table 8.1 for each trial.

5. Since only m_1 was moving before and only m_2 was moving after the collision, the momentum values of the two carts represent the total momentum before and after the collision. Calculate and record in Data Table 8.1 the percent difference in the total momentum before and after the collision for each trial. Identify and record the *best run* from the three trials.

Part B: Unequal Mass Collision

1. Attach about 400 grams of additional mass to a cart, which we will call m_3. Record the mass of m_1 and m_3 in Data Table 8.2 on page 93. Place m_3 at rest near the center position of the track. Place m_1 near the end of the track.

2. Conduct a trial run by giving m_1 an initial velocity toward m_3 at rest. As before, the m_1 cart should have its initial velocity *before* reaching the first reference mark tape (if this method is being used). After the collision, m_1 should be moving in an opposite direction from its initial velocity and m_3 should be in motion. This will require the m_1 timing team to measure the velocity both before *and* after the collision. (For a spark timer, turn the timer on for the first half of the tape length only, then turn it back on after the collision to make marks on the other half.)

3. Conduct three trials of giving m_1 an initial velocity toward m_3 at rest. Record the velocity before and after each collision and record in Data Table 8.2. Calculate the momentum of m_1 before the collision and after the collision. Calculate the momentum of m_3 after each collision. Record these calculations in Data Table 8.2 for each trial.

4. Calculate and record in Data Table 8.2 the percent difference in the total momentum before (which is m_1 only) and after the collision for each trial.

Results

1. Describe the possible sources of error in this experiment.

2. Was linear momentum conserved according to the results of this experiment (what did you expect and what did you find)? What is the evidence that supports your response?

3. At the moment that two carts collide, they both experience an instantaneous stop. Is momentum conserved during this stop? Give reasons for your answer.

4. What solid evidence can you provide that the cart was *not* accelerating?

5. Was the purpose of this lab accomplished? Why or why not? (Your answer to this question should be reasonable and make sense, showing thoughtful analysis and careful, thorough thinking.)

Invitation to Inquiry

How are momentum changes and collision forces related? A rubber ball, a ball of modeling clay, and a super ball all exert forces, and all undergo a momentum change when they strike the floor or a wall. Do they all exert the same force if they have the same momentum? Do they all undergo the same momentum change? Set up a demonstration of a pendulum with a rubber ball, ball of modeling clay, and a super ball as a bob. If all have the same mass, will all move a wooden block over the same distance if released from the same height? Or, is it necessary to adjust the height to see the same results? Note the clay might stick to the block, the super ball will bounce back at almost the same speed, and the rubber ball will be somewhere between sticking and bouncing back with the same speed. Make your predictions, then take a survey of your classmates about their predictions. Can anyone give an explanation for their predictions in terms of Newton's second law of motion, momentum, and the force exerted by moving objects? Do the demonstration to find answers to this question.

	m_1 before Collision			m_2 after Collision			Difference
Trial	Mass (g)	Velocity (cm/s)	Momentum (p)	Mass (g)	Velocity (cm/s)	Momentum (p)	Percent (%)
1							
2							
3							

Data Table 8.1   Momentum of Equal Mass Collision

Best run momentum before: _____   Best run momentum after: _____

Best run percent difference for momentum before and momentum after: _____

Space for Calculations:

Data Table 8.2   Momentum of Unequal Mass Collision

Trial	m_1 before collision			Place for Calculations:
	Mass (g)	Velocity (cm/s)	Momentum (p_1)	$p_1 = p_2 - p_1$
1				
2				
3				

Trial	m_1 after collision			m_3 after collison			
	Mass (g)	Velocity (cm/s)	Momentum $(-p_1)$	Mass (g)	Velocity (cm/s)	Momentum (p_3)	Percent Difference (%)
1							
2							
3							

Experiment 9: Rotational Equilibrium

Introduction

In addition to straight line, projectile, and motion around a circular path, an object can spin, or rotate about an axis, and this movement is called **rotational motion**. One way to measure how fast something is turning is to count the number of revolutions that occur during a given time. Depending on the nature of what is doing the turning, the frequency of rotation can be expressed in revolutions per second (rps), revolutions per minute (rpm), or revolutions per hour (rph).

What is required to produce rotational motion? If you have ever tried to loosen a stuck bolt with a wrench, you know that where you grasp the wrench, that is, the distance from your hand to the bolt is important. The direction you push is also important, of course, as is the magnitude of the force. These are factors that affect the **torque**, or *tendency of a force to produce a rotation*. As you can see in figure 9.1, applying a torque is the same as applying a force perpendicular to a lever. A lever arm is defined as the perpendicular distance from a center of rotation to an applied force vector. With everything else equal, a force applied farthest from the center of rotation has the longest lever arm and produces the greatest torque (figure 9.1). The force applied at right angles (90°) to the lever arm, as opposed to some other angle, also produces the greatest torque. It is often useful to think of a torque as a product obtained by multiplying a force by a lever arm length, or Torque = force × lever arm. In symbols the relationship is

$$\tau = Fr$$

where τ (the Greek letter *tau*) is the torque, F is the applied force, and r is the perpendicular distance from the center of rotation. Torque units are usually the newton-meter (foot-pound). Torque units are similar to units of energy or work, but they represent completely different concepts.

It is torque, and not force that should be considered for rotational motion. Two equal but opposite parallel forces on an object would not produce a translation motion because the net force would be zero. It is possible, however, for two equal but opposite parallel forces to result in a net

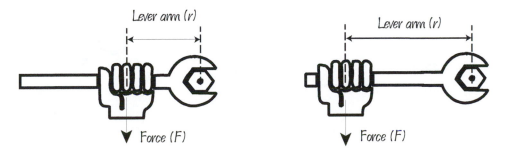

Figure 9.1

force of zero, but produce at the same time a net torque that is greater than zero. For example, a lever with the pivot point in the center will turn if two equal but opposite forces are applied in the same rotational direction. The lever rotates because the torques are greater than zero. There is no translation motion in this situation because the net force is zero.

Newton's laws of motion apply to rotational motion as well as translational motion, but in the case of rotational motion the torque, not the force is considered. Thus for the first law of motion you can see that an object that is not rotating will continue not to rotate as long as the net torque is zero. Likewise, an object that is rotating uniformly will continue to do so at a constant speed as long as no net torque acts to change the motion. In this experiment you will study torques that are needed to keep a body that is not rotating from rotating, that is, torques in rotational equilibrium.

Procedure

A set-up for this experiment could be the apparatus as shown in figure 9.2. It shows a meterstick with a knife-edge clamp on a support stand and masses hanging from movable mass hangers. If this apparatus is not available, a meter stick could be balanced by a loop of string suspended from a support (such as a second meter stick placed on top of two chair backs), with the string tied and held in place with masking tape. The masses could also be suspended by loops of string around the meter stick, which could also be held in place with masking tape. This experiment could be conducted with a wide variety of setups, and the following general procedure can be used. In any case, the first step is to *balance the meterstick*. Record the balance point, which will be called x_0, in Data Table 9.1 on page 101.

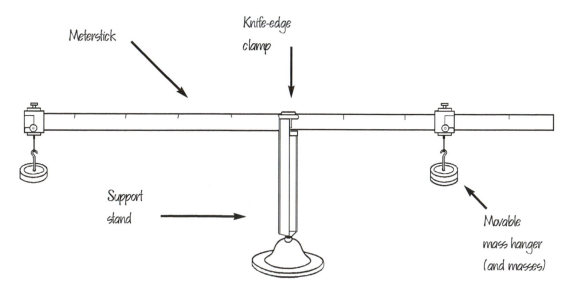

Figure 9.2

Part A: Two Unequal Masses in Equilibrium

1. Hang a 100 g mass (m_1) at a position 50 cm from x_0 on either side of the meterstick. Record the mass of m_1 and its position as x_1 in Data Table 9.1. Note the position of x_1 is the distance from x_0. Also include the mass of any clamp and hanger that may be used with m_1.

2. Hang a 200 g mass (m_2) on the opposite side of the meterstick from m_1. Adjust the position of m_2 until static equilibrium is found. Record the mass of m_2 and its position as x_2 in Data Table 9.1. Again, note the position of x_2 is the distance from x_0 and also be sure to include the mass of any clamp and hanger that may be used with m_2.

3. Calculate the torque τ_1 around x_0 from the force of gravity on the 100 g mass, $\tau_1 = m_1 g x_1$, where g is 9.8 m/s^2, the average acceleration of gravity. Show your calculations here and record τ_1 in Data Table 9.1.

4. Calculate the torque τ_2 around x_0 from the force of gravity on the 200 g mass, $\tau_2 = m_2 g x_2$, where g is 9.8 m/s^2. Show your calculations here and record τ_2 in Data Table 9.1.

5. Compare the clockwise and counterclockwise torques on a sketch in Data Table 9.1 and record the percent difference. Record your calculations and comments, if any, here:

Part B: Three Unequal Masses in Equilibrium

1. Hang a 50 g mass (m_1) and a 100 g mass (m_2) at two different positions *on the same side* of the meter stick and record the position of x_1 from x_0 and the position of x_2 from x_0 in Data Table 9.2 on page 102. Also include the mass of any clamp and hanger that may be used with m_1 and m_2.

2. Hang a 200 g mass (m_3) on the opposite side of the meterstick from m_1.and m_2. Adjust the position of m_3 until static equilibrium is found. Record the mass of m_3 and its position as x_3 in Data Table 9.2. Again, be sure to include the mass of any clamp and hanger that may be used with m_3.

3. Calculate the torque τ_1 around x_0. Show your calculations here and record τ_1 in Data Table 9.2.

4. Calculate the torque τ_2 around x_0. Show your calculations here and record τ_2 in Data Table 9.2.

5. Calculate the torque τ_3 around x_0. Show your calculations here and record τ_3 in Data Table 9.2.

6. Compare the clockwise and counterclockwise torques on a sketch in Data Table 9.2 and record the percent difference between ($\tau_1 + \tau_2$) and τ_3. Record your calculations and comments, if any, here:

Part C: Unknown Mass in Equilibrium

1. Hang an unknown mass (m_1) on one side of the meter stick, but near x_0, and record the position of x_1 from x_0 in Data Table 9.3 on page 103.

2. Hang a 90 g mass (m_2) on the opposite side of the meterstick from m_1. Adjust the position of m_2 until static equilibrium is found. Record the mass of m_2 and its position as x_2 in Data Table 9.3. Include the mass of any clamp and hanger that may be used with m_2.

3. Calculate and compare the torque τ_2 from the mass m_2 and distance x_2 to the torque τ_1 from the unknown mass and the known distance x_1. Use the equation $\tau_1 = \tau_2$ to solve for the unknown mass m_1. Show your calculations here and record your findings in Data Table 9.3.

4. Use a laboratory balance to find the mass of m_1, recording the measurement in Data Table 9.3.

5. Compare the value of the mass of m_1 as determined from the clockwise and counterclockwise torques to the value of the mass of m_1 as determined by the laboratory balance. Calculate the percent difference between these two methods in Data Table 9.3. Record your calculations and comments here:

Results

1. What is a torque? Why are torques, not just forces, considered for rotational equilibrium?

2. Explain why masses are moved back and forth, from notch to notch, along a scale on a laboratory balance.

3. Compare the measurement errors that might result from determining an unknown mass as in Part C of this experiment with the measurement errors that might result from using a laboratory balance to determine the mass.

4. Was the purpose of this lab accomplished? Why or why not? (Your answer to this question should be reasonable and make sense, showing thoughtful analysis and careful, thorough thinking.)

Invitation to Inquiry

As you may have noticed, a moving bicycle is more stable than a stationary one and it may be difficult to stay upright on the bike when stopped. Investigate the relationship between torque and spinning with a bicycle wheel mounted on an axle. Find a way to measure the torque needed to change the axis of rotation, say from a vertical to a horizontal direction. Compare the torque needed to do this for a slowly spinning wheel as compared to a rapidly spinning one. Also compare any differences observed when you apply the torque quickly in a fast flip as compared to application of torque slowly in a slow flip. What relationships were found?

Data Table 9.1   Two Unequal Masses in Equilibrium

Balance point (x_0)-- _____

Mass (m_1)-- _____

Distance of x_1 from x_0-- _____

Torque τ_1 -- _____

Mass (m_2)-- _____

Distance of x_2 from x_0-- _____

Torque τ_2 -- _____

Percent difference in τ_1 and τ_2 --------------------------------- _____

Sketch showing labeled clockwise and counterclockwise torques:

Data Table 9.2   Three Unequal Masses in Equilibrium

Mass (m_1)-- _____

Distance of x_1 from x_0-- _____

Torque τ_1 --- _____

Mass (m_2)-- _____

Distance of x_2 from x_0-- _____

Torque τ_2--- _____

Mass (m_3)-- _____

Distance of x_3 from x_0-- _____

Torque τ_3--- _____

Percent difference in $\tau_1 + \tau_2$ and τ_3 -------------------------- _____

Sketch showing labeled clockwise and counterclockwise torques:

Data Table 9.3 Unknown Mass in Equilibrium

Distance of x_1 from x_0 --- _____

Torque τ_1 --- _____

Mass (m_2) --- _____

Distance of x_2 from x_0 --- _____

Torque τ_2 --- _____

Calculated m_1 --- _____

Measured m_1 on laboratory balance ------------------------ _____

Percent difference --- _____

Sketch showing labeled clockwise and counterclockwise torques:

Experiment 10: Centripetal Force

Introduction

This experiment is concerned with the force necessary to keep an object moving in a constant circular path. According to Newton's first law of motion there *must be* forces acting on an object moving in a circular path since it does not move off in a straight line. The second law of motion ($F = ma$) also indicates forces since an unbalanced force is required to change the motion of an object. An object moving in a circular path is continuously being accelerated since it is continuously changing direction. This means that there is a continuous unbalanced force acting on the object that pulls it out of a straight-line path. The force that pulls an object out of a straight-line path and into a circular path is called a **centripetal force.**

The magnitude of the centripetal force required to keep an object in a circular path depends on the inertia (or mass) and the acceleration of the object, as you know from the second law ($F = ma$). The acceleration of an object moving in uniform circular motion is $a = v^2/r$, so the magnitude of the centripetal force of an object with a mass (m) that is moving with a velocity (v) in a circular orbit of radius (r) can be found from

$$F = \frac{mv^2}{r}$$

The distance (circumference) around a circle is $2\pi r$. The velocity of an object moving in a circular path can be found from $v = d/t$, or $v = 2\pi r/T$ where $2\pi r$ is the distance around one complete circle and T is the period (time) required to make one revolution. Substituting for v,

$$F = \frac{m\left(\dfrac{2\pi r}{T}\right)^2}{r}$$

or

$$F = \frac{\dfrac{m4\pi^2 r^2}{T^2}}{r},$$

$$F = \frac{4\pi^2 r^2 m}{T^2} \times \frac{1}{r}$$

$$F = \frac{4\pi^2 r\, m}{T^2}$$

This is the relationship between the centripetal force (F_c), the mass (m) of the object in circular motion, the radius (r) of the circle, and the time (T) required for one complete revolution.

Procedure

1. The equipment setup for this experiment consists of weights (washers) attached to a string, and a rubber stopper that swings in a horizontal circle. You will swing the stopper in a circle and adjust the speed so that the stopper does not have a tendency to move in or out, thus balancing the centripetal force (F_c) on the stopper with the balancing force (F_b), or $m_w g$, exerted by the washers on the string.

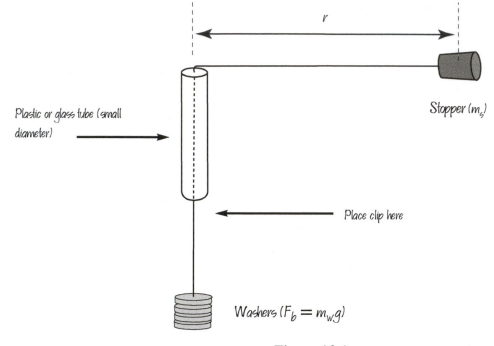

Figure 10.1

2. Place some washers on the string and practice rotating the stopper by placing a finger next to the string, then moving your hand in a circular motion. You are trying to move the stopper with a consistent, balancing motion, just enough so the stopper does not move in or out. *Keep the stopper moving in a fairly horizontal circle, without the washers moving up or down.* An alligator (or paper) clip placed on the string just below the tube will help you maintain a consistent motion by providing a point of reference as well as helping with length measurements. Be careful of the moving stopper so it does not hit you in the head.

3. After you have learned to move the stopper with a constant motion in a horizontal plane, you are ready to take measurements. The distance from the string at the top of the tube to the *center* of the stopper is the radius (r) of the circle of rotation. The mass (m_s) of the stopper is determined with a balance. The balancing force (F_b) of the washers is determined from the mass of the washers times g ($F_b = m_w g$). The period (T) is determined by measuring the time of a number of revolutions, then dividing the total time by the number of revolutions to obtain the time for one revolution. For example, 20 revolutions in 10 seconds would mean that $^{10}/_{20}$, or 0.5 seconds, is required for one revolution. This data is best obtained by one person acting as a counter speaking aloud while another person acts as a timer.

4. Make four or five trials by rotating the stopper with a different number of washers on the string each time, adding or removing two washers (about 20 g) for each trial. For each trial, record in Data Table 10.1 the mass of the washers, the radius of the circle, and the average time for a single revolution.

Data Table 10.1	Centripetal Force Relationships				
Trial	Mass of washers (m_w)	Balancing force (F_b)	Radius (r)	Time (t)	Centripetal force (F_c)
	(kg)	(N)	(m)	(s)	(N)
1	_____	_____	_____	_____	_____
2	_____	_____	_____	_____	_____
3	_____	_____	_____	_____	_____
4	_____	_____	_____	_____	_____
5	_____	_____	_____	_____	_____

Mass of stopper (m_s) _____ kg

5. Calculate and record the balancing force (F_b) for each trial from the mass of washers times g (9.8 m/s^2), or $F_b = m_w g$.

6. Calculate and record the centripetal force (F_c) for each trial from

$$F_c = \frac{4\pi^2 r\, m_s}{T^2}$$

Considering the balancing force (F_b) as the accepted value, and the calculated centripetal force (F_c) as the experimental value, calculate your percentage error for each trial of this experiment. Analyze the percentage errors and other variables to identify some trends, if any.

Trial 1 :

Trial 2 :

Trial 3 :

Trial 4 :

Trial 5 :

Results

1. Did the balancing force (F_b) equal the centripetal force (F_c)? Do you consider them equal or not equal? Why or why not?

2. Analyze the errors that could be made in all the measured quantities. What was probably the greatest source of error and why? Discuss how these errors could be avoided and how the experiment in general could be improved.

3. Discuss any trends that were noted in your analysis of percentage error for the different trials. Analyze the meaning of any observed trends or discuss the meaning of the lack of any trends.

4. Was the purpose of this lab accomplished? Why or why not? (Your answer to this question should be reasonable and make sense, showing thoughtful analysis and careful, thorough thinking.)

Invitation to Inquiry

1. Did you ever try to figure out which is a cooked egg and which is a raw one without breaking the shell? One way to accomplish this is by spinning the eggs on a plate, and the well-cooked one will continue to spin while the uncooked egg will rock back and forth. The yolk is heavier than the white, but why would an uncooked egg spin more slowly? Use your understanding of centripetal force to develop some ideas about why eggs should behave this way, then design a demonstration or experiment to test your idea.

2. Experiment with some things that rotate, such as rolling cylinders. Roll large, small, solid, hollow, and various combinations of large and solid cylinders, small and solid cylinders down an incline. Predict ahead of time which will reach the bottom of the incline first. Then test your predictions.

3. A hollow and solid cylinder of the same size do not have the same weight. If you roll the two cylinders down an incline slope together, side by side, the solid cylinder should win. Yet if you attach strings of equal lengths to make pendulums from the same hollow and solid cylinders, then you will find that they swing together, side by side. Is this true? Experiment to find out, then be prepared to explain your findings.

4. Explore relationships between mass distance from an axis and how hard it is to set an object into rotational motion. Consider using a baton with some kind of movable masses that can be fixed to the baton different distances from the axis of rotation. A large wooden dowel rod and lumps of clay might be a good experimental alternative to a baton.

Experiment 11: Archimedes' Principle

Introduction

Archimedes' principle states that an object floating or submerged in a liquid is buoyed up by a force equal to the weight of the liquid displaced by the object. The purpose of this experiment is to study this principle as it applies to floating and submerged objects and its application to the determination of specific gravity.

The buoyant force on an object immersed in a liquid can be determined by weighing an object in air and then in water. The apparent loss of weight in water, or $W_w - W_a$, where W_a is the weight in air and W_w is the weight in water, is the buoyant force of the water. The weight of the water displaced by an object can be measured by using an overflow can and catch bucket (figure 11.1). The relationship between the buoyant force on an immersed object can then be compared to the weight of the water displaced.

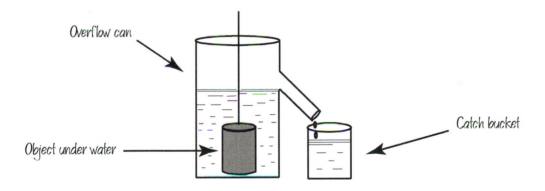

Figure 11.1

The specific gravity of an object is defined as the ratio of the density of the object to the density of water at the same temperature. According to Archimedes' principle, the apparent loss of weight of an object immersed in a liquid is equal to the weight of the liquid displaced. The specific gravity of an object more dense than water is easily determined by weighing the object in air, then weighing it suspended in water. The loss of weight in water is equal to the weight of the water displaced, that is, the weight of an equal volume of water. The loss of weight in water is $W_w - W_a$, where W_a is the weight in air and W_w is the weight in water, so the specific gravity (S) is equal to

$$S = \frac{W_a}{W_a - W_w}$$

Procedure

1. Weigh the object in air and record the weight in Data Table 11.1.

2. Fill an overflow can with water while holding a finger over the spout hole. Place the can on the table, with the spout hole over a catch bucket. Remove your finger from the spout hole, catching the excess water. When the water in the can is level with the spout, discard the excess water from the catch bucket and dry it with paper towels. Record the weight of the dry catch bucket in Data Table 11.1. Place the dry catch bucket beneath the spout hole again.

3. Weigh the object in water by attaching it with a fine thread that is attached to a spring scale or laboratory balance, then lowering it into the overflow can. Immerse the object completely, catching the overflow water in the catch bucket as shown in figure 11.1. Record the weight in Data Table 11.1.

4. Complete the calculations in Data Table 11.1.

5. Repeat procedure steps 1-4, this time with a floating object such as a block of wood. Record the measurements and calculations in Data Table 11.2.

Results

1. Compare the weight of the *submerged object,* the buoyant force on the object, and the weight of the water displaced by the object:

2. Compare the weight of the *floating object,* the buoyant force on the object, and the weight of the water displaced by the object:

3. What generalization would describe your comparisons in questions 1 and 2?

4. What is the specific gravity of the submerged object according to the measurements in Data Table 11.1, as compared to the known value (if available)?

5. Was the purpose of this lab accomplished? Why or why not? (Your answer to this question should show thoughtful analysis and careful, thorough thinking.)

Invitation to Inquiry

1. Explain why a ball of clay sinks, but when shaped into a boat, it floats. Experiment with a ball of clay and container of water to find which shape of clay will hold the most "cargo."

2. Try to float an egg in a glass with a measured amount of fresh water. Keeping track of how much you use, add salt to the water, stirring, until the egg floats. Calculate the density of the egg from the salt and water measurements. Then calculate the density of the egg by weighing and finding the volume by water displacement.

3. Make a hydrometer by adding modeling clay to the bottom of a plastic soda straw until it floats upright. Mark the water level for when the hydrometer is floating in distilled water, then compare to the level in other liquids. Visit a service station or shop where antifreeze and car batteries are checked. Find out how specific gravity is used to make these tests.

Data Table 11.1	Buoyant force on Submerged Object
1. Weight of object in air	_____
2. Apparent weight of object in water	_____
3. Buoyant force on object (row 1 minus row 2)	_____
4. Weight of empty catch bucket	_____
5. Weight of catch bucket plus overflow water	_____
6. Weight of displaced water (row 4 minus row 5)	_____

Data Table 11.2 Buoyant force on Floating Object	
1. Weight of object in air	_____
2. Apparent weight of object in water	_____
3. Buoyant force on object (row 1 minus row 2)	_____
4. Weight of empty catch bucket	_____
5. Weight of catch bucket plus overflow water	_____
6. Weight of displaced water (row 4 minus row 5)	_____

Experiment 12: Boyle's Law

Introduction

A gas is not very dense, compared to solids and liquids, and it diffuses rapidly throughout any container in which it is placed. This diffusion is spontaneous and against gravity and only the movement of highly mobile molecules can account for it. The gas molecules must be rapidly moving about, colliding often with other gas molecules and the walls of the container. Since it is not necessary to supply heat to maintain the temperature of a perfectly insulated sample of gas, all of these collisions must be perfectly plastic. On the average, there is no loss of kinetic energy when molecules collide.

The assumptions you can logically make about gases can be extended to explain some of the observable properties of gases. For example, the pressure (force per unit area) that a gas exerts on a surface can be assumed to be the result of the continuous bombardment of gas molecules on the surface. This would explain why adding more air to a tire increases the air pressure in the tire since more air means more bombardment by the added molecules. Likewise, the air pressure increases in your tires when you drive because friction with the road increases the absolute temperature of the air in your tires. Increasing the absolute temperature increases the average molecular kinetic energy, which may increase the frequency of impact on the tire walls as well as the force of impact. In any event, increased frequency and increased force of impact will result in greater air pressure.

In this experiment you will investigate how the volume of a given mass of gas varies with the pressure exerted on it. The pressure and volume of a fixed amount of gas at constant temperature are inversely proportional, which is the relationship known as Boyle's law. In symbols, Boyle's law is

$$P = \frac{k}{V} \quad \text{or} \quad PV = k$$

where P is the gas pressure, V is the volume of the gas, and k is a proportionality constant with a value that depends on the constant temperature and mass of the gas. Suppose V_1 is used to represent the initial volume and V_2 a new volume after a pressure change. Then P_1 and P_2 can be used to represent the initial and new pressures. With the same temperature and mass of gas between the initial and new volume and pressure ($k = k$), then

$$\frac{V_1}{V_2} = \frac{P_1}{P_2} \quad \text{or} \quad P_1V_1 = P_2V_2$$

The experimental test of Boyle's law consists of observing a series of volumes, measuring the corresponding pressures, and plotting P versus $1/V$ to see if a straight line is obtained.

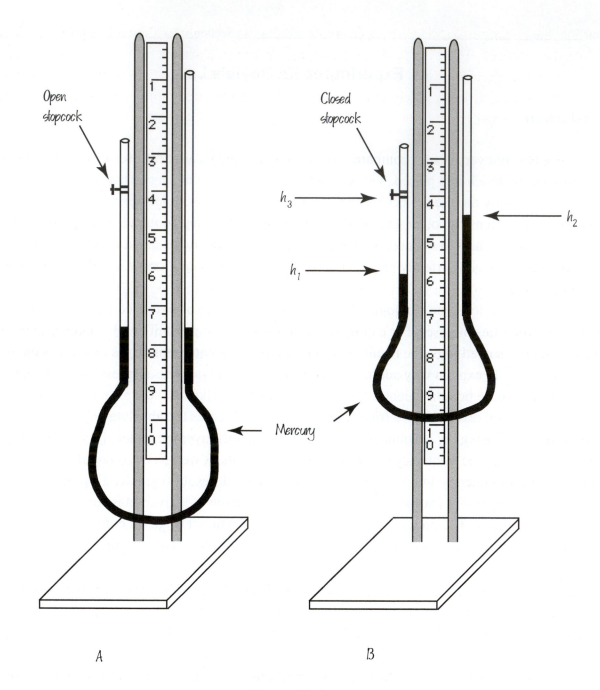

Open
stopcock

Closed
stopcock

h_3

h_2

h_1

Mercury

A

B

Figure 12.1

Procedure

1. Measure the barometric pressure and record the reading in Data Table 12.1.

2. Inspect the Boyle's law apparatus to make sure that each tube is about half full when the stopcock is open. **CAUTION:** Inform your laboratory instructor of any mercury spills, no matter how small. Any spilled mercury should be cleaned up and properly disposed of before it becomes "lost" in the environment.

3. Adjust the apparatus so the mercury columns are the same height as shown in figure 12.1A, then close the stopcock. The closed side of the apparatus (the side with the stopcock) will remain stationary and should not be touched or handled during the investigation. Note that the mercury columns remain the same height after the stopcock is closed, meaning that atmospheric pressure exists in the closed volume.

4. Measure the top of the closed tube just below the stopcock and record this measurement in Data Table 12.1 as h_3.

5. Move the open side of the tube down as far as you can. This decreases the pressure on the air in the closed tube and the difference in the mercury levels of the two tubes, h_2 and h_1, is the gauge pressure. Record the measurements of h_2 and h_1 in Data Table 12.1, then calculate and record the gauge pressure. Do not move the open tube until after the next procedure step.

6. Since the glass tube has a uniform bore, the volume of air inside the tube is proportional to the length of the tube. The length of the column of air inside the closed tube, $h_3 - h_1$, is therefore representative of the volume of air inside the tube. Calculate the volume of air inside the closed tube and record it in Data Table 12.1.

7. Raise the open tube to the highest position possible in eleven approximately equal steps, recording the h_2 and h_1 for each of the steps. Calculate and record the gauge pressure and the volume of air inside the tube for each of the 12 different observations, recording all calculations in Data Table 12.1.

Results

1. Calculate the total pressure on the confined air in each of the 12 different observations, recording your findings in Data Table 12.1. **Note:** If gauge pressure is indicated in cm, record the barometric pressure also in cm to calculate the total pressure in Data Table 12.1.

2. Calculate *PV* for each of the 12 different observations, again recording your findings in Data Table 12.1.

3. Calculate 1/V for each of the 12 observations and record your findings in the data table.

4. Plot the total pressure versus the corresponding values of 1/V. Calculate the slope and write it on the graph somewhere and here as well:

5. Describe how the calculations in Data Table 12.1 would verify Boyle's law.

6. Describe how your graph would verify Boyle's law.

7. Does the *PV* column of Data Table 12.1 agree with your graph? Explain in detail, giving evidence with your answer.

8. Explain how touching or handling the closed tube could introduce error into this experiment. What are other sources of error?

9. Was the purpose of this lab accomplished? Why or why not? (Your answer to this question should show thoughtful analysis and careful, thorough thinking.)

Invitation to Inquiry

Make a Cartesian diver by putting a little water in a small test tube, then inverting it in an almost filled, small-mouth bottle of water. Adjust the water inside the small tube so it just floats, then place a rubber stopper in the bottle. Pressure on the bottle will cause the inverted tube to sink because pressure compresses the air inside the little tube, allowing in more water which increases its weight. If you can find a flat (flask-shaped) bottle, you can adjust the tube so pushing on the flat sides makes it sink and stay down, but pushing the other way makes it rise again. Use the diver to quantify some volume and pressure relationships.

Data Table 12.1   Boyle's Law

Trial	h_1	h_2	Gauge Pressure $(h_2 - h_1)$	Volume $(h_3 - h_1)$	Total Pressure (Atmospheric plus Gauge)	PV	$\dfrac{1}{V}$
1	_____	_____	_____	_____	_____	_____	_____
2	_____	_____	_____	_____	_____	_____	_____
3	_____	_____	_____	_____	_____	_____	_____
4	_____	_____	_____	_____	_____	_____	_____
5	_____	_____	_____	_____	_____	_____	_____
6	_____	_____	_____	_____	_____	_____	_____

Barometer Reading _____

Measurement at Top of Closed Tube (h_3) _____

Data Table 12.1   Boyle's Law, continued

Trial	h_1	h_2	Gauge Pressure $(h_2 - h_1)$	Volume $(h_3 - h_1)$	Total Pressure (atmospheric plus Gauge)	PV	$\frac{1}{V}$
7	———	———	———	———	———	———	———
8	———	———	———	———	———	———	———
9	———	———	———	———	———	———	———
10	———	———	———	———	———	———	———
11	———	———	———	———	———	———	———
12	———	———	———	———	———	———	———

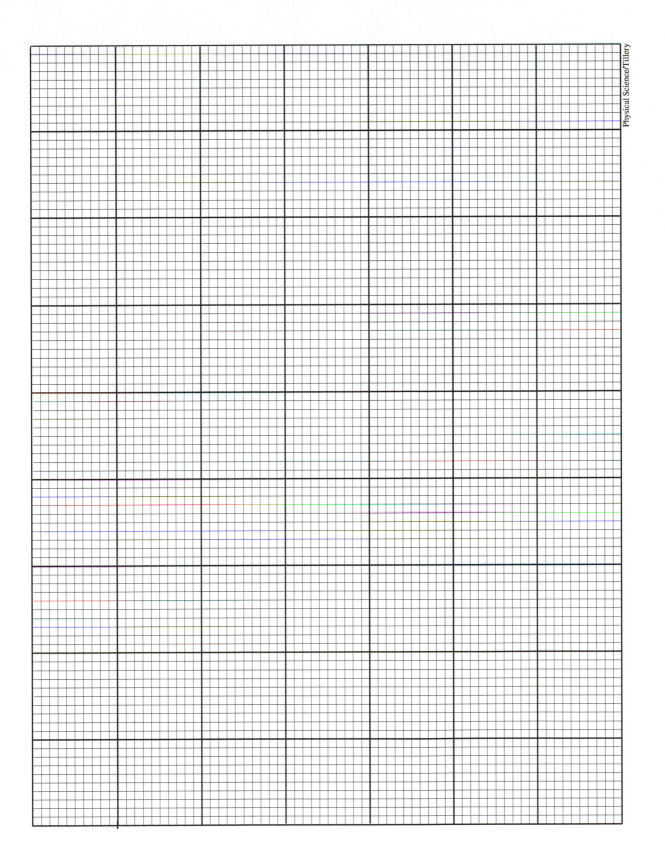

Experiment 13: Work and Power

Introduction

The word *work* represents a concept that has a special meaning in science that is somewhat different from your everyday concept of the term. In science, the concept of work is concerned with the application of a force to an object and the distance the object moves as a result of the force. **Work** (*W*) is defined as the magnitude of the applied force (*F*) multiplied by the distance (*d*) through which the force acts, $W = Fd$.

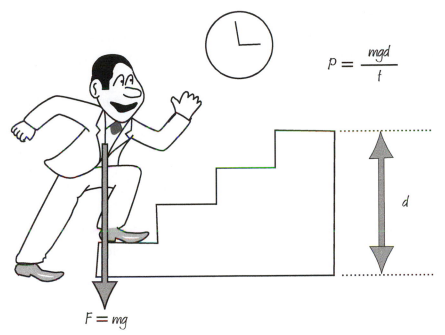

$$P = \frac{mgd}{t}$$

$$F = mg$$

Figure 13.1

You are doing work when you walk up a stairway since you are lifting yourself through a distance. You are lifting your weight (the force exerted) the vertical height of the stairs (distance through which the force is exerted). Running up the stairs rather than walking is more tiring because you use up your energy at a greater rate when running. The rate at which energy is transformed or the rate at which work is done is called power. **Power** (*P*) is defined as work (*W*) per unit of time (*t*),

$$P = \frac{W}{t}$$

When the steam engine was first invented there was a need to describe the rate at which the engine could do work. Since people at that time were familiar with using horses to do their work, the steam engines were compared to horses. James Watt, who designed a workable steam engine, defined

horsepower (hp) as a power rating of 550 ft·lb/s. In SI units, power is measured in joules per second, called the **watt** (W). It takes 746 W to equal 1 hp, and 1 kW is equal to about 1⅓ hp.

Procedure

1. Teams of two volunteers will measure the work done, the rate at which work is done, and the horsepower rating as they move up a stairwell. Person A will measure and record the data for person B. Person B will measure and record the data for person A. An ordinary bathroom scale can be used to measure each person's weight. Record the weight in pounds (lb) in Data Table 13.1. This weight is the force (F) needed by each person to lift himself or herself up the stairs.

2. The vertical height of the stairs can be found by measuring the height of one step, then multiplying by the number of steps in the stairs. Record this distance (d) in feet (ft) in Data Table 13.1.

3. Measure and record the time required for each person to *walk normally* up the flight of stairs. Record the time in seconds (s) in Data Table 13.1.

4. Measure and record the time required for each person to *run* up the flight of stairs as fast as can be safely accomplished. Record the time in seconds (s) in Data Table 13.1.

5. Calculate the work accomplished, power level developed, and horsepower of each person while walking and while running up the flight of steps. Be sure to include the correct units when recording the results in Data Table 13.1.

Results

1. Explain why there is a difference in the horsepower developed in walking and running up the flight of stairs.

2. Is there some limit to the height of the flight of stairs used and the horsepower developed? Explain.

3. Could the horsepower developed by a slower-moving student ever be greater than the horsepower developed by a faster-moving student? Explain.

4. Describe an experiment that you could do to measure the horsepower you could develop for a long period of time rather than for a short burst up a stairwell.

5. Was the purpose of this lab accomplished? Why or why not? (Your answer to this question should show thoughtful analysis and careful, thorough thinking.)

Invitation to Inquiry

You may have seen a person use a bare hand to break a board with a Karate chop. The person hits the 3/4 inch (1.9 cm) board with a quick movement, transferring enough kinetic energy to break the board. How much kinetic energy is needed to break the board? You can find this if you know how much work is needed to break it. Obtain a 15 by 30 cm pine board, cut so the grain is parallel to the shorter side. Figure out a way to hang weights around the center of the board as you measure how much the board bends under the increasing weight. When it breaks, you will have the force needed and the distance the force moved the board, so you will be able to calculate the work done. Working backward from the work needed, you will be able to find the kinetic energy you need to transfer to break the board.

	Volunteer A		Volunteer B	
Data Table 13.1 — Work and Power Data and Calculations				
	Walking	Running	Walking	Running
Weight (F) (lb)	_____	_____	_____	_____
Vertical height (d) of steps (ft)	_____	_____	_____	_____
Time required (t) to *walk* the flight of steps (s)	_____		_____	
Time required (t) to *run* the flight of steps (s)		_____		_____
Work done $W = Fd$	_____	_____	_____	_____
Power $P = W/t$	_____	_____	_____	_____
Horsepower developed $P \div 550$ ft·lb/s	_____	_____	_____	_____

Experiment 14: Friction

Introduction

Friction is the resisting force that opposes the sliding or rolling motion of one body over another. The force needed to overcome friction depends on the nature of the materials in contact, their smoothness, and on the normal force. Imagine a large crate on a horizontal floor. You want to move it by pushing it to the right with a horizontal force, F (figure 14.1). If F is small, the frictional force (f) acting to the left will keep the crate from moving ($F = f$) and the crate will remain in equilibrium. Since the crate is not moving, this frictional force is called the **force of static friction** (f_s). If the magnitude of F is gradually increased, F will eventually exceed f_s and the crate will move to the right. Once you start the crate moving, you will find that it takes the force F' to keep it moving at a constant velocity. The crate is in equilibrium so there must be no unbalanced force on it and the force tending to stop the motion must be equal to and opposite to the force F'. The retarding frictional force for an object in motion is called the **force of kinetic friction** (f_k). Since the crate moves with a constant speed, then you know that $F' = f_k$. If F' were to increase as to become unbalanced, the crate would accelerate to the right. If F' were to decrease to make f_k larger, the crate would accelerate to rest. Both f_s and f_k are proportional to the normal force and both depend on the nature of the surfaces in contact. This relationship can be summarized by the following equation,

$$F = \mu N$$

where F is the applied force opposing friction and μ (the Greek letter *mu*) is called the **coefficient of friction**. As you can see, μ is the fraction of the normal force that it takes to make surfaces slide over

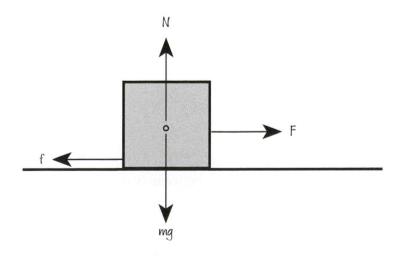

Figure 14.1

129

each other. In this particular example the normal force is the weight but this is not always the case. Normal means perpendicular, *not vertical*. So, we can write the equilibrium condition as

$$\Sigma F = 0$$

and so

$$0 = F - |\mu N|$$

or

$$F = |\mu N|.$$

The absolute value signs are around the term with the N because it is perpendicular to the direction of F so one cannot legitimately set $F = N$ times a constant. Two vectors cannot be equal if they are not parallel, but this symbol is often not written. The coefficient of friction that opposes the direction of motion is called the coefficient of sliding friction or the **coefficient of kinetic friction**. It is the one used when the two surfaces are moving relative to one another and often has the symbol μ_k. The coefficient of kinetic friction could also be expressed as a ratio of the force of kinetic friction to the normal force produced by two surfaces pressing together, or

$$\mu_k = \frac{F}{N}$$

where F is the force of kinetic friction directed parallel to the surfaces and opposite to the direction of motion, N is the normal force, and μ_k is the coefficient of kinetic friction. In cases where the two surfaces are *not moving* relative to one another we use the **coefficient of static friction**, μ_s. This coefficient can be expressed as a ratio of the force of static friction (F_s) to the normal force necessary to start movement, or

$$\mu_s = \frac{F_s}{N}$$

In this investigation, you will use these relationships to determine the coefficient of static friction and the coefficient of kinetic friction between two wood surfaces.

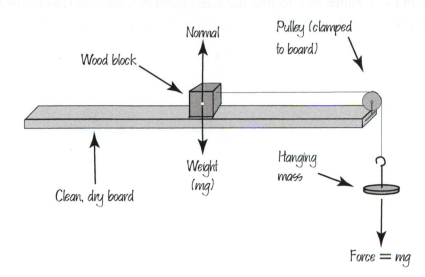

Figure 14.2

Procedure

1. Use a balance to find the mass of the wood block. Record the mass in Data Table 14.1 on page 133, then calculate and record the weight of the block.

2. Place a clean, dry board on the laboratory table with its pulley over the edge of the table. Place the block on the board with the largest area in contact with the board. A light cord is attached to the block and a weight hanger, then run over the pulley. Add small masses to the hanger until the force created is just sufficient to keep the block moving slowly with a constant speed after it has been started with a gentle push. Record this force (*mg*) in Data Table 14.1.

3. Repeat procedure step 2, this time placing increasing masses on top of the block and recording the force needed to keep the block moving slowly with a constant speed when it has been started with a gentle push. Record in Data Table 14.1 the mass on the block along with the force needed to keep the block moving at a constant speed.

4. Place a 500 gram mass (or masses) on top of the block. Gradually increase the mass of the hanger until the block moves, without a push, with a uniform speed. If the block accelerates, start over and use slightly less mass on the hanger. Do this three times, and average the results. Record the average force needed (*mg*) in Data Table 14.2 on page 134.

5. Use the information in Data Table 14.1 to plot the values of the force of friction vs. the values of the normal force (see figure 14.1). Calculate the slope to obtain the coefficient of kinetic friction μ_k for wood on wood. Write the value here and somewhere on the graph.

6. Use the information in Data Table 14.2 to find the coefficient of static friction μ_s for wood on wood. Use the equation

$$\mu_s = \frac{F_s}{N}$$

and show your calculations here.

Results

1. Which was greater, the coefficient of static friction or the coefficient of kinetic friction? Is this the result you were expecting? Explain.

2. Why was it necessary for the block to move with a constant velocity in all procedures?

3. How consistent were the friction effects observed in procedure step 4? Why would this be the case?

4. Was the purpose of this lab accomplished? Why or why not? (Your answer to this question should be reasonable and make sense, showing thoughtful analysis and careful, thorough thinking.)

Data Table 14.1   Coefficient of Kinetic Friction

Mass Placed on Block (m) (kg)	Mass of Block (m) (kg)	Total Mass (m) (kg)	Total Weight of Block and Masses (mg) (N)	Total Normal Force (N)	Force Needed to Move Block Uniformly (N)
_____	_____	_____	_____	_____	_____
_____	_____	_____	_____	_____	_____
_____	_____	_____	_____	_____	_____
_____	_____	_____	_____	_____	_____
_____	_____	_____	_____	_____	_____
_____	_____	_____	_____	_____	_____

Coefficient of kinetic friction (from graph) _____

Data Table 14.2	Coefficient of Static Friction			
Mass Placed on Block	Weight Placed on Block	Total Normal Force	Force Required to Move Block	Coefficient of Static Friction
Trial 1 _____	_____	_____	_____	_____
Trial 2 _____	_____	_____	_____	_____
Trial 3 _____	_____	_____	_____	_____
Average _____	_____	_____	_____	_____

Invitation to Inquiry

1. How would changing the area of contact between the block and board affect (a) the μ_k and (b) the μ_s? Make your prediction below, then experiment with the one of the smaller surfaces of the block to test your prediction.

2. What would happen to the frictional force and the coefficient of friction if the board is raised at one end? What happens to the frictional force and the coefficient of friction as the angle is increased more and more?

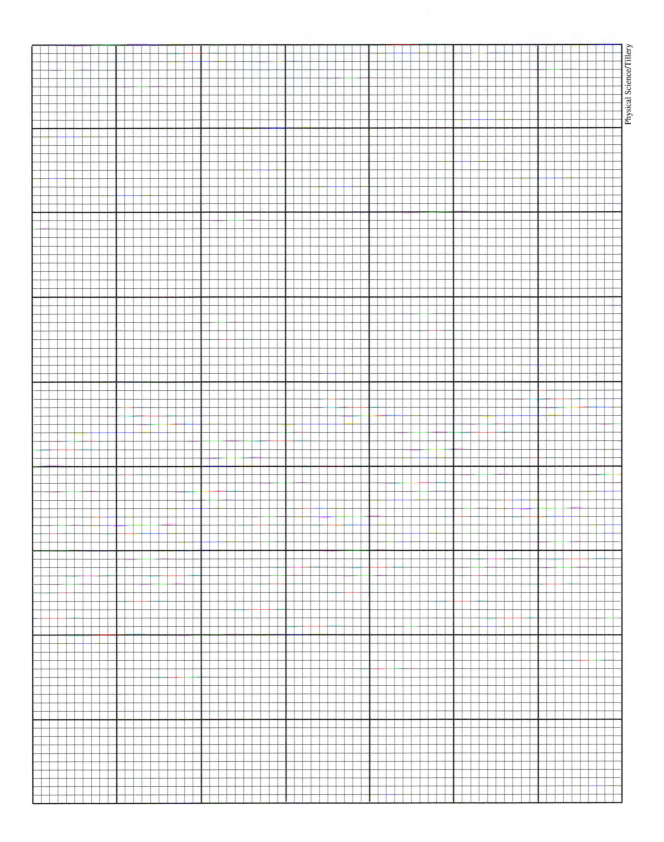

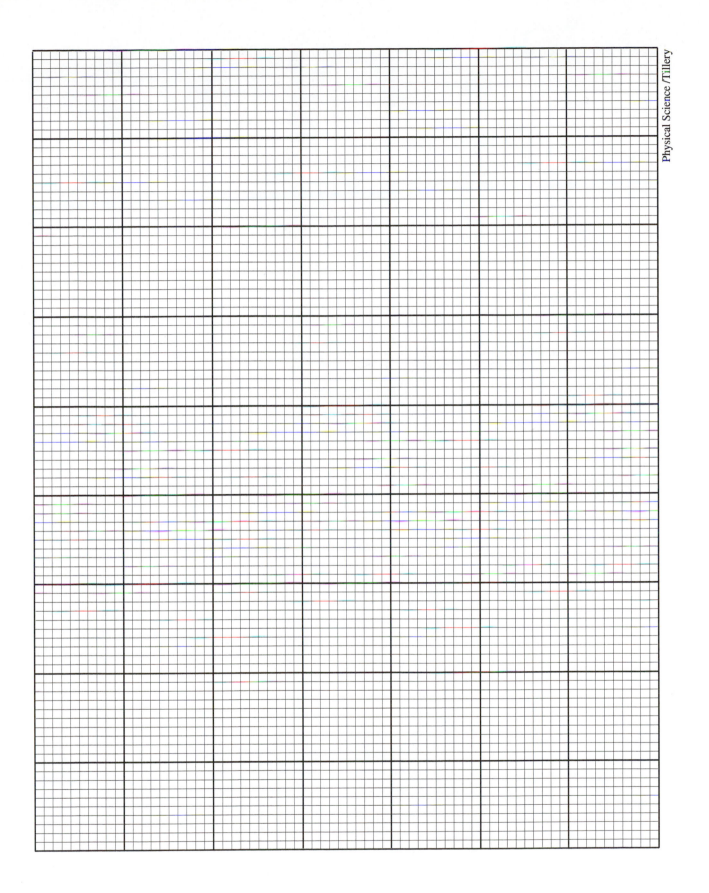

Experiment 15: Hooke's Law

Introduction

Considering what happens to a solid when it is squeezed or stretched leads to the conclusion that molecules have strong forces of interaction. Some materials are deformed, or "bent out of shape" by squeezing or stretching forces, but return to their original shape when the squeezing or stretching force is removed. Some materials do this better than other ones and the materials that do it best have the property of elasticity. Elastic materials return to their original shape after being deformed by some external force—as long as the force was not too great. If you press gently on the exterior metal side of a car door, you can see the metal return to its original shape when you stop pressing—if you didn't press too hard. Elastic materials such as the metal of your car door have intermolecular forces that pull the molecules back to their original positions. This pulls the solid as a whole back to its original shape when the applied force is removed. Such responses to compression or stretching forces lead to the assumption that molecules repel one another if pushed closer together and attract one another if pulled further apart.

There is a limit to how much external force a given material can experience and still have the intermolecular forces pull or push the material back to its original shape. The elastic limit is the maximum force per unit area that a material can be subjected to before becoming permanently deformed. At the elastic limit the intermolecular forces are overcome and the molecules slide past one another, permanently altering the shape of the object. If enough force is applied the material will break or fracture. At the fracture point the intermolecular forces are overcome to such an extent that the molecules are completely separated. It is the property of elasticity that causes a rubber ball to bounce. The force of striking the floor deforms the ball when it is dropped. As the ball regains its original shape, it pushes itself away from the floor causing the ball to rise above the floor, or bounce. A clay ball does not bounce because it exhibits plastic deformation. Plastic is the opposite of elastic. A material does not recover from a plastic deformation and stays in the shape into which it was deformed. The materials called plastics were so named because they exhibit this property when warm. Both elastic and plastic deformations take place before the fracture point is reached.

When a coiled spring is stretched by a weight, it is deformed but will return to its original shape and length. If the spring is stretched too far, however, the elastic limit may be exceeded and the spring will be permanently stretched out of shape. Before the elastic limit is reached, a spring will stretch by a length that is proportional to the weight pulling on it. For example, if a one pound weight stretches the spring one inch, a two pound weight will stretch it two inches, and a three pound weight will stretch it three inches. You could say that the deformation is directly proportional to the applied force within the elastic limit. The directly proportional relationship between the stretch, or change in length ΔL, and the applied force F is called **Hooke's Law**. In symbols,

$$F = -\Delta kL$$

where k is a proportionality constant depending on the elastic material. Note that k is positive; the sign is negative because the internal force exerted by the elastic materials is in a direction opposite to the applied force, that is, the force exerted by the spring and ΔL are in opposite directions. The larger k is, the greater force (F) needed to cause a given ΔL.

The proportionality constant, k, may be measured for any object by applying a known force, F, to cause a measurable deformation, ΔL. The restoring force will be equal in magnitude to the applied force, and k can thus be determined by using

$$k = \frac{F}{\Delta L}$$

As you can see, k will have units of N/m. Solved for F, and with ΔL expressed as the change of length from the original length ($L_f - L_i$), the equation becomes

$$F = k\,(L_f - L_i)$$

or

$$F = (kL_f - kL_i).$$

Note this equation is in the same form as the equation for a straight line, which has a general form of

$$y = mx + b,$$

where m is the slope and b is the y-intercept.

Procedure

1. Hang a spring, small end up, to a rigid support. Place a 50 g weight hanger on the lower end. Initially, enough weight must be placed on the hanger so that no two sections of the spring touch each other and all the kinks are out of the spring. The length of the straightened spring and the attached weight will be used as a starting point for all subsequent length and weight measurements. Measure this length from the top of the spring to the bottom of the weight hanger and record the length in Data Table 15.1 on page 143.

2. Add 0.1 kg to the weight hanger. Measure and record the new length to the bottom of the weight hanger.

3. Take a total of ten more sets of measurements covering a total distance of at least 0.1 m. Record all measurements in the data table and complete the necessary calculations.

Results

1. Make a graph of each data point, with the total distance on the *x*-axis and the total weight on the *y*-axis. Calculate the slope, including units, and write the value of the slope here and also somewhere on the graph.

2. Use the data to calculate an average k from $mg/\Delta L$ and discuss why this value should or should not be equal to the slope obtained in question 1.

3. Use the results from your calculation in question 2 and the value of the slope calculated in question 1 to determine a percent difference. What were the percentage difference and probable sources of experimental error?

4. Discuss how you could improve the precision of this experiment.

5. Was the purpose of this lab accomplished? Why or why not? (Your answer to this question should show thoughtful analysis and careful, thorough thinking.)

Invitation to Inquiry

Using the same spring and set-up of this experiment, adjust the total mass on the spring to 500 g. Pull the spring down 5 cm and release it. Measure its period by timing 5 to 10 oscillations or cycles. Divide the total time by the number of cycles to find the average period. The equation for the period of a spring is

$$T = 2\pi\sqrt{\frac{m}{k}}$$

where $\pi = 3.1416$, m = mass, and k = the spring constant. Use this equation to determine the spring constant k, then compare it to the value found from the spring elongation graph in the experiment.

Data Table 15.1   Hooke's Law

Mass of straightened spring and hanger masses _____ kg

Weight of straightened spring and hanger masses _____ N

Length of spring and weight hanger _____ m

Length of Spring	Total Mass	Total Weight	k
_____ m	_____ kg	_____ N	_____ N/m
_____ m	_____ kg	_____ N	_____ N/m
_____ m	_____ kg	_____ N	_____ N/m
_____ m	_____ kg	_____ N	_____ N/m
_____ m	_____ kg	_____ N	_____ N/m
_____ m	_____ kg	_____ N	_____ N/m
_____ m	_____ kg	_____ N	_____ N/m
_____ m	_____ kg	_____ N	_____ N/m
_____ m	_____ kg	_____ N	_____ N/m
_____ m	_____ kg	_____ N	_____ N/m
_____ m	_____ kg	_____ N	_____ N/m

Average k _____ N/m

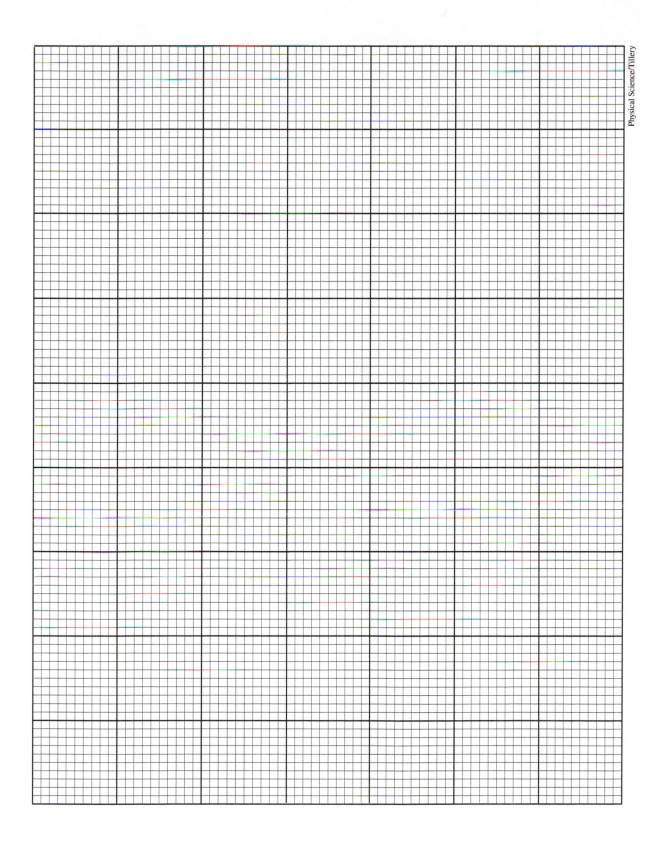

Experiment 16: Thermometer Fixed Points

Introduction

This experiment is concerned with the fixed reference points on the Fahrenheit (T_F) and Celsius (T_C) thermometer scales. Two easily reproducible temperatures are used for the fixed reference points and the same points are used to define both scales. The fixed points are the temperature of melting ice and the temperature of boiling water under normal atmospheric pressure. The differences in the two scales are (1) the numbers assigned to the fixed points, and (2) the number of divisions, called **degrees**, between the two points. On the Fahrenheit scale, the value of 32 is assigned to the lower fixed point and the value of 212 is assigned to the upper fixed point, with 180 divisions between these two points. On the Celsius scale, the value of 0 is assigned to the lower fixed point and the value of 100 is assigned to the upper fixed point, with 100 divisions between these two points. In this laboratory investigation you will compare observed thermometer readings with the actual true fixed points.

Variations in atmospheric pressure have a negligible effect on the melting point of ice but have a significant effect on the boiling point of water. Water boils at a higher temperature when the atmospheric pressure is greater than normal, and at a lower temperature when the atmospheric pressure is less than normal. Normal atmospheric pressure, also called **standard barometric pressure**, is defined as the atmospheric pressure that will support a 760 mm column of mercury. An atmospheric pressure change that increases the height of the column of mercury will increase the boiling point by 0.037°C (0.067°F) for each 1.0 mm of additional height. Likewise, an atmospheric pressure change that decreases the height of the column will decrease the boiling point by 0.037°C (0.067°F) for each 1.0 mm of decreased height. Thus you should add 0.037C° for each 1.0 mm of a laboratory barometer reading above 760 mm and subtract 0.037C° for each 1.0 mm below the normal pressure of 760 mm. This calculation will give you the actual boiling point of water under current atmospheric pressure conditions. Any difference between this value and the observed thermometer reading is an error in the thermometer.

Procedure

1. First, verify accuracy of the lower fixed point of the thermometer. Fill a beaker with cracked ice as shown in figure 16.1. After water begins forming from melting ice, place the bulb end of the thermometer well into the ice, but leave the lower fixed point on the scale uncovered so you can still read it. Gently stir for five minutes and then until you observe no downward movement of the mercury. When you are confident that the mercury has reached its lowest point, carefully read the temperature. The last digit of your reading should be an estimate of the distance between the smallest marked divisions on the scale. Record this observed temperature of the melting point in

Data Table 16.1. Use 0°C as the accepted value and calculate and record the measurement error, if any.

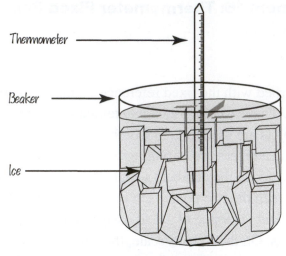

Figure 16.1

2. Now verify the accuracy of the upper fixed point of the thermometer. Set up the steam generator as illustrated in figure 16.2. If you need to insert the thermometer in the stopper, be sure to moisten both with soapy water first. Then hold the stopper with a cloth around your hand and *gently* move the thermometer with a twisting motion. The water in the steam generator should be adjusted so the water level is about 1 cm below the thermometer bulb. When the water begins to boil vigorously, observe the mercury level until you are confident that it has reached its highest point. Again, the last digit of your reading should be an estimate of the distance between the smallest

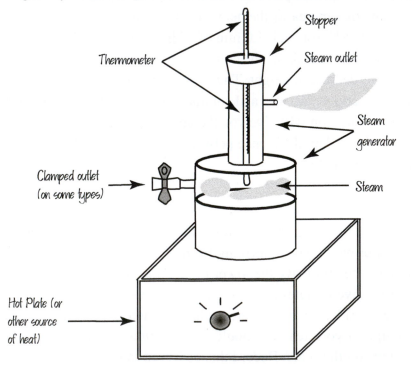

Figure 16.2

marked divisions on the thermometer scale. Record this observed temperature of the boiling point in Data Table 16.1.

3. Determine the accepted value for the boiling point by recording in mm the barometric pressure, then calculating the deviation above or below 100°C. Record this accepted boiling point in Data Table 16.1, then calculate and record the measurement error, if any.

4. Repeat the entire procedure for a second trial, recording all data in Data Table 16.1.

Results

1. Did the temperature change while the ice was melting? Offer an explanation for this observation.

2. Describe how changes in the atmospheric pressure affect the boiling point of water. Offer an explanation for this relationship.

3. Account for any differences observed in the melting point and boiling point readings.

4. How would the differences determined in this investigation influence an experiment concerning temperature if the errors were not considered?

5. Was the purpose of this lab accomplished? Why or why not? (Your answer to this question should show thoughtful analysis and careful, thorough thinking.)

Going Further

Using data from your *best* trial, make a graph by plotting the Celsius temperature scale on the x-axis and the Fahrenheit temperature scale on the y-axis. Calculate the slope of the straight line and write it here and on the graph somewhere, then answer the following questions:

1. What is the value of the slope? What is the meaning of the slope?

2. What is the value of the y-intercept?

3. The slope-intercept form for the equation of a line is $y = mx + b$, where y is the variable on the y-axis (in this case, °F), x is the variable on the x-axis (in this case, °C), m is the slope of the line, and b is the y-intercept. Use this information to write the equation of the Celsius-Fahrenheit temperature graph. What is the meaning of this equation?

Invitation to Inquiry

1. Wash an aluminum pop can, leaving a small amount of water in the can. Use tongs to hold the can over a heat source until the water boils, and you can see steam condensing in the air at the opening. Immediately invert the can part way in a container of cool water. Explain what happens in terms of Charles' law, Boyle's law, and a molecular point of view.

2. Place the ends of a one meter metal rod on two wood blocks and secure one end to its block. Place a pin through a tagboard pointer under the free end. The metal rod should be able to move back and forth, turning the pin as it moves. Explain what happens to the pointer as the metal rod is heated, then cooled.

3. Boil a small amount of water in a clean 500 mL flask, then apply a round balloon over the mouth before the flask cools. Place a rubber band, doubled if necessary, around the balloon on the neck of the flask. Explain what happens to the balloon as the flask is heated or cooled without using the terms "drawn in" or "suck."

Data Table 16.1	Thermometer Readings and Actual Fixed Points	
	Trial 1	Trial 2
Observed melting point	_____	_____
Measurement error— melting ice	_____	_____
Observed boiling point	_____	_____
Barometric pressure	_____	_____
Deviation from normal (+ or –)	_____	_____
Accepted boiling point	_____	_____
Measurement error— boiling water	_____	_____

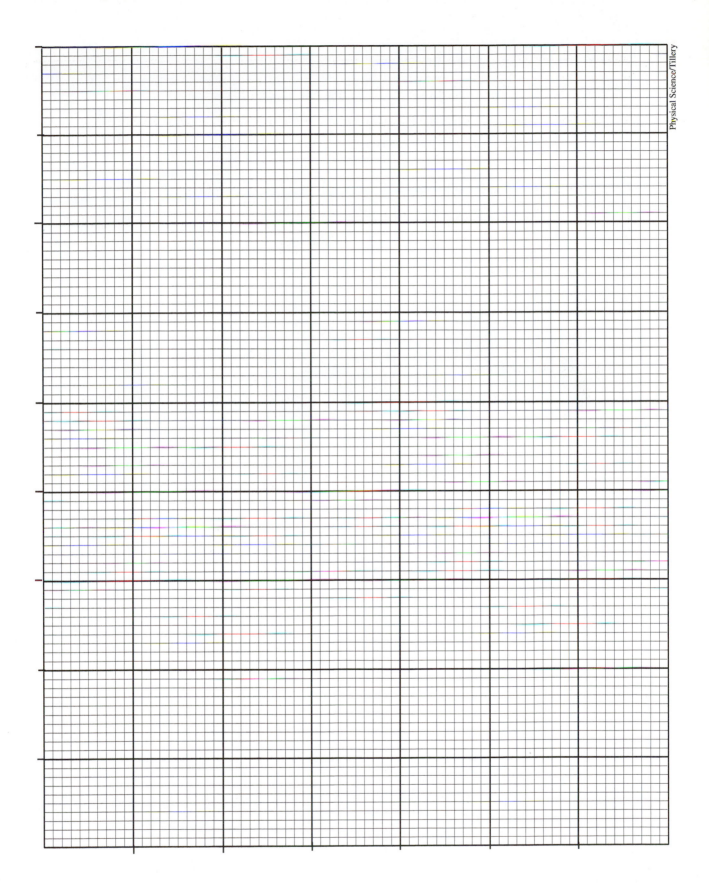

Experiment 17: Absolute Zero

Introduction

If you place an inflated balloon in a freezer, you will find that the balloon becomes much smaller as it cools. Warming the balloon back to its original temperature results in an increased volume, which will be the original volume if no air leaked during the cooling and warming process. The volume of air in the balloon depends on the temperature of the air and increases or decreases in the volume are proportional to increases or decreases in the temperature. The change in volume is proportional to the change in temperature, or $V \propto T$.

Since the volume of a gas decreases with decreases in temperature, how much of a temperature decrease would be needed to decrease the volume of the gas to zero? A liter of gas loses about 1/273 of its volume for each degree it is cooled from 0°C. So, you might project a zero volume occurring at a temperature of –273°C. This will not happen, of course, since the gas will liquefy and then perhaps solidify as the temperature is lowered. There are many other relationships, however, that point to –273°C as a special temperature. It is the coldest temperature possible, the complete lack of heat, and it is called **absolute zero**. The Kelvin scale has the same degree interval size as the Celsius scale but begins at absolute zero. The Kelvin scale is more than a Celsius scale with the zero point moved down by 273 degrees. The Kelvin scale is absolute, not relative to some arbitrary fixed points as are the Celsius and Fahrenheit scales. Thus, zero on the Kelvin scale *does* mean zero, and all the numbers above zero have meaning in relation to one another. This relationship does not exist on the relative number scales.

Thus if you double the absolute temperature of a gas its volume doubles as $V \propto T$. If the temperature remains constant, the volume of the gas is inversely proportional to the pressure, or

$$V \propto \frac{1}{P}$$

Combining these two relationships,
By inserting a proportionality constant k, you can write the relationship as an equation that is known

$$V \propto \frac{T}{P} \quad \text{or} \quad PV \propto T$$

as the **ideal-gas law**,

You will use the ideal-gas law in this experiment. You will observe a small glass tube that is

$$PV = kT \quad \text{where } T \text{ is in Kelvins.}$$

closed at one end and has a movable bead of mercury near the other end. The air pressure inside the tube increases or decreases with changes in temperature. The atmospheric pressure is constant, so the mercury bead moves back and forth with the changing pressure of the air trapped inside. Because the small glass tube is of uniform diameter, at a constant pressure the length of the air column should be proportional to its volume, which should in turn be proportional to the temperature.

155

Procedure

1. The small glass tube with the mercury bead is placed inside a larger tube so hot and cold fluids can flow evenly and completely around the small tube, uniformly changing the temperature of the air inside the tube (figure 17.1).

2. Your laboratory instructor will adjust the position of the mercury bead so it is about two-thirds of the total length from the closed end.

3. You will measure the length of the column of air inside the small tube by measuring from the closed end of the tube to the inside edge of the mercury bead. Measure the bead at a right angle to the length of the tube in all trials. It is also important to measure the length of the column of air from the side of the bead that is closest to the closed end of the tube. The mercury will expand and contract with changes in temperature, as well as change positions with the column of air.

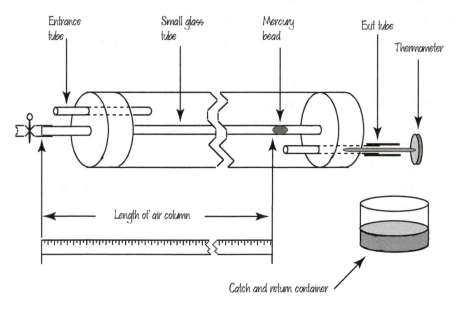

Figure 17.1

4. Measure the length of the air column inside the small tube at room temperature. Record this length and the room temperature in Data Table 17.1.

5. Adjust the large tube so the exit end is slightly higher than the entrance end (figure 17.2). The tube should slope slightly upwards, allowing air bubbles to leave the tube as you run water through it. Run ice water through the large tube, catching and returning the water with a catch container, then pouring it back over the funnel full of ice. Have a second container ready to catch water while you are pouring water back through the funnel. Continue recycling the water through the ice until the temperature at the exit is constant. Record this temperature and the length of the air column in Data Table 17.1.

156

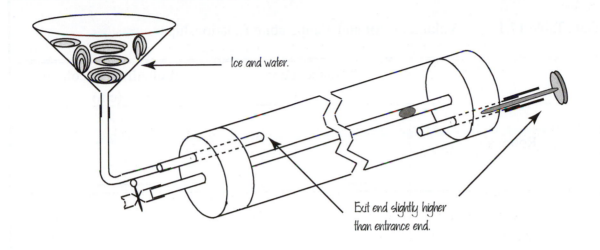

Ice and water.

Exit end slightly higher than entrance end.

Figure 17.2

6. With the tube still sloping slightly upwards to allow the escape of air bubbles, run hot water through the system until the temperature at the exit is constant. Record this temperature and the length of the air column in Data Table 17.1.

7. Adjust the large tube so it slopes slightly downward (figure 17.3). Run steam through the system until the temperature at the exit is constant. The downward slope will permit water that condenses from the steam to escape. Record the temperature and the length of the air column in Data Table 17.1.

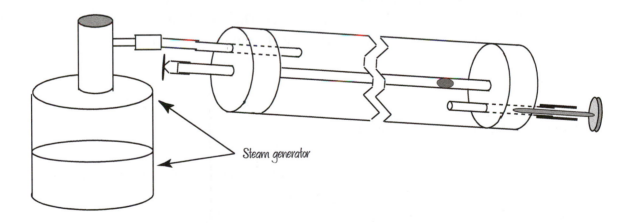

Steam generator

Figure 17.3

Data Table 17.1	Volume of Air and Temperature Relationships	
	Temperature (T_c)	Length of air column (cm)
Room air	_____	_____
Ice water	_____	_____
Hot water	_____	_____
Steam	_____	_____

Results

1. Analysis by extrapolation: Make a graph with the *x*-axis running from −350°C to +120°C. Plot the data points and make a straight best-fit line in the region of the data points. Using a ruler, continue with a dotted line until it crosses the *x*-axis. The dotted line will cross the *x*-axis at the projected, or extrapolated, Celsius temperature of absolute temperature.

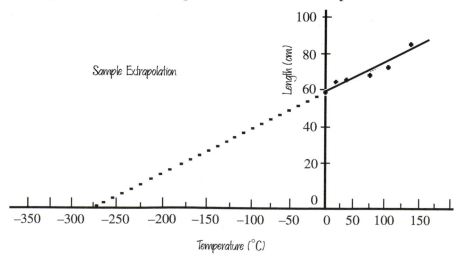

2. Analysis from slope: Make a second graph with the x-axis running from $0°$ C to $100°$ C. Plot the data points and draw a best-fit line. Determine the slope of the line. Determine the y-intercept. Use the slope-intercept equation of a straight line, $y = mx + b$, to find x when $y = 0$. Since y is the length of the air column and x is the temperature, then

$$0 = (\text{slope})(\text{temperature}) + b$$

or

$$\text{temperature}\left(0\ \text{K}\right) = \frac{-b}{\text{slope}}$$

3. Find the experimental error of analysis by extrapolation, using $-273.15°C$ as the accepted value of absolute zero.

4. Find the experimental error of analysis by equation of slope, using $-273.15°C$ as the accepted value of absolute zero.

5. Was the purpose of this lab accomplished? Why or why not? (Your answer to this question should show thoughtful analysis and careful, thorough thinking.)

Invitation to Inquiry

The sketch below is of a termometro lento, an interesting device first named by Galileo Galilei (1564-1642). It is a "floating glass bulb" thermometer and the name means "slow thermometer." This thermometer is "slow" because it depends on density changes that occur with changes in the temperature. In general, the density of a liquid increases with decreases in temperature and this increases the buoyancy of the liquid. All of the bulbs have different masses of liquids in same-volume bulbs, so each bulb has its own density. When the temperature of the liquid in this thermometer rises, it becomes less dense and the bulbs will sink slowly one by one, according to their density. In the same way, the bulbs are buoyed up as the temperature decreases. The temperature is read by the lowest floating bulb (see sketch below). The density of each bulb has been calibrated in a temperature-stabilized bath, and this temperature is marked on the bulb.

Is the termometro lento drawn below in a cool or warm environment? What would you need to know to build such a thermometer? How much would you need to vary the mass in each bulb to create a thermometer with a two degree Fahrenheit scale?

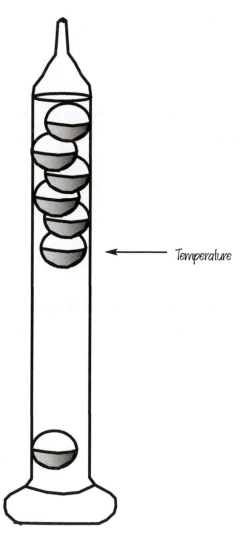

Temperature

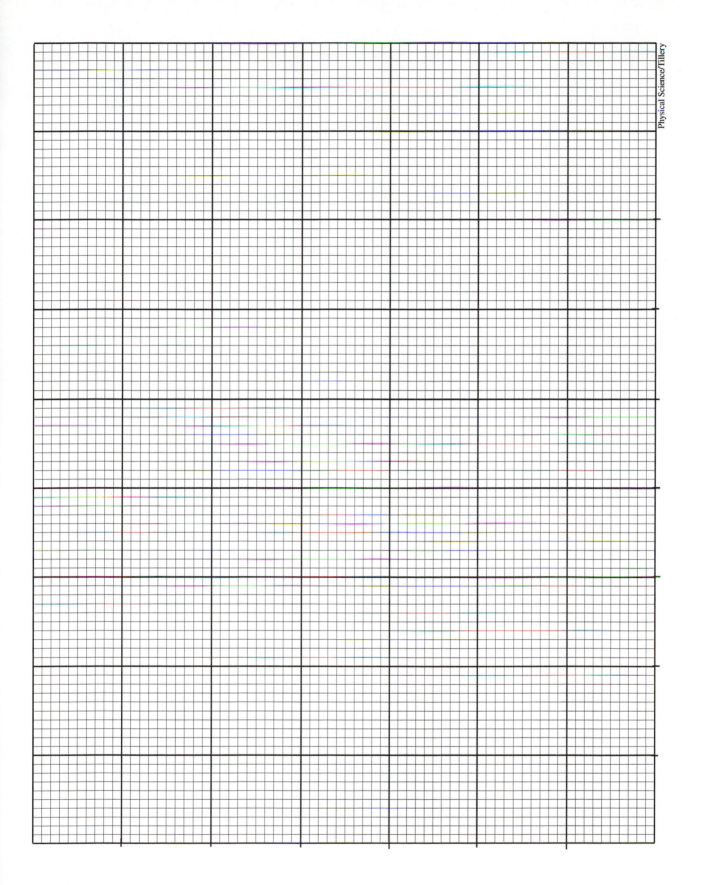

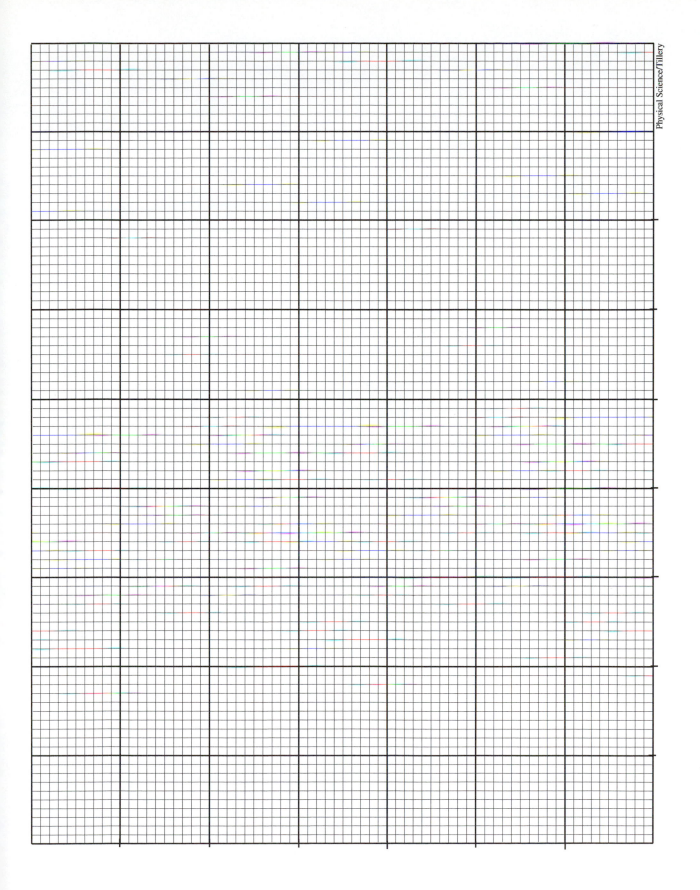

Experiment 18: Specific Heat

Introduction

Heating is a result of energy transfer, and a quantity of heat can be measured just as any other quantity of energy. The metric unit for measuring energy or heat is the **joule**. However, the separate historical development of the concepts of motion and energy and the concepts of heat resulted in separate units. Some of these units are based on temperature differences.

The metric unit of heat is called the **calorie** (cal), a leftover term from the old caloric theory of heat. A calorie is defined as the amount of energy (or heat) needed to increase the temperature of one gram of water one degree Celsius. A kilocalorie (kcal) is the amount of energy (or heat) needed to increase the temperature of one kilogram of water one degree Celsius. The relationship between joules and calories is called the **mechanical equivalence of heat,** and the relationship is

$$4.184 \text{ J} = 1 \text{ cal}$$
or
$$4184 \text{ J} = 1 \text{ kcal}.$$

There are three variables that influence the energy transfer that takes place during heating: (1) the temperature change, (2) the mass of the substance being heated, and (3) the nature of the material being heated. The relationships among these variables are:

1. The quantity of heat (Q) needed to increase the temperature of a substance from an initial temperature of T_i to a final temperature of T_f is proportional to $T_f - T_i$, or $Q \propto \Delta T$.
2. The quantity of heat (Q) absorbed or given off during a certain ΔT is also proportional to the mass (m) of the substance being heated or cooled, or $Q \propto m$.
3. Differences in the nature of materials result in different quantities of heat (Q) being required to heat equal masses of different substances through the same temperature range.

The **specific heat** (c) is the amount of energy (or heat) needed to increase the temperature of one gram of a substance one degree Celsius. The property of specific heat describes the amount of heat required to heat a certain mass through a certain temperature change, so the units for specific heat are cal/g°C or kcal/kg°C. Note that the k's in the second set of units cancel, so the numerical value for both is the same — for example, the specific heat of aluminum is 0.217 cal/g°C, or 0.217 kcal/kg°C. Some examples of specific heats in these units are:

Aluminum 0.217 cal/g°C   Iron 0.113 cal/g°C
Copper 0.093 cal/g°C   Silver 0.056 cal/g°C
Lead 0.031 cal/g°C   Nickel 0.106 cal/g°C

When the units of all three sets of relationships are the same units used to measure Q, then all the relationships can be combined in equation form,

$$Q = mc\Delta T.$$

This relationship can be used for problems of heating or cooling. A negative result means that energy is leaving a material; that is, the material is cooling. When two materials of different temperatures are involved in heat transfer and are perfectly insulated from their surroundings, the heat lost by one will equal the heat gained by the other,

$$\text{Heat lost}_{\text{(by warm substance)}} = \text{Heat gained}_{\text{(by cool substance)}}$$
$$\text{or}$$
$$Q_{\text{lost}} = Q_{\text{gained}}$$
$$\text{or}$$
$$(mc\Delta T)_{\text{lost}} = (mc\Delta T)_{\text{gained}}.$$

Calorimetry consists of using the concept of conservation of energy and applying it to a mixture of materials initially at different temperatures that come to a common temperature. In other words,

$$\text{(heat lost by sample)} = \text{(heat gained by water)}.$$

The sample is heated, then placed in water in a calorimeter cup where it loses heat. The water is initially cool, gaining heat when the warmer sample is added. (The role of a Styrofoam calorimeter cup in the heat transfer process can be ignored since two Styrofoam cups have negligible heat gain [$\Delta T \approx 0$] and very little mass.) In symbols,

$$m_s c_s \Delta T_s \quad = \quad m_w c_w \Delta T_w$$

where m_s is the mass of the sample, c_s the specific heat of the sample, and ΔT_s is the temperature change for the sample. The same symbols with a subscript w are used for the mass, specific heat, and temperature change of the water. Solving for the specific heat of the sample gives

$$c_s \quad = \quad \frac{m_w c_w \Delta T_w}{m_s \Delta T_s}$$

Procedure

1. You are going to determine the specific heat of three samples of different metals by using calorimetry. You will run two trials on each sample, making *very careful* temperature and mass measurements. Do the calculations before you leave the lab. If you have made a mistake you will still have time to repeat the measurements if you know this before you leave.

2. Be sure you have sufficient water to cover at least the bottom two-thirds of a submerged metal boiler cup (see figure 18.1), but not so much water that it could slosh into the cup when the water is boiling. Start heating the water to a full boil as you proceed to the next steps.

166

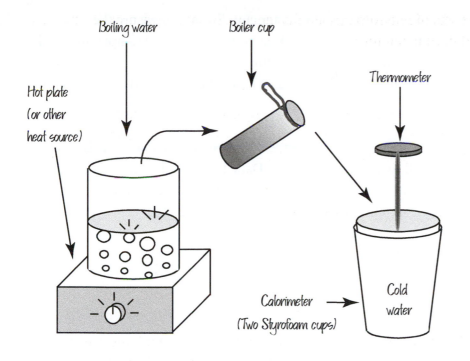

Figure 18.1

3. Measure and record the mass of a dry boiler cup. Pour metal shot into the boiler cup until it is about one-third filled, then measure and record the mass of the cup plus shot. Record the mass of the metal sample (m_s) in Data Table 18.1.

4. Carefully insert a thermometer into the metal shot, positioning it so the sensing end is in the middle of the shot, not touching the sides of the boiler cup. Carefully lower the boiler cup into the boiling water. Heat the metal shot until it is in the range of 90° to 95°C. Allow the sample to continue heating as you prepare the water and calorimeter cup (steps 5 and 6).

5. Acquire or make a calorimeter cup of two Styrofoam cups, one placed inside the other (figure 18.1) to increase the insulating ability of the cup. Measure and record the mass of the two cups. Add just enough water to the cup to cover the metal shot when it is added to the cup. This water should be cooler than room temperature (this is to balance possible heat loss by radiation). Measure and record the initial temperature of the water (T_{iw}) in Data Table 18.1.

6. Determine the mass of the cup with the water in it, then subtract the mass of the cup to find the mass of the cold water (m_w). Record the mass of the cold water in Data Table 18.1.

7. Measure and record the temperature of the metal shot. Record the initial temperature of the sample (T_{is}) in Data Table 18.1.

8. Pour the metal shot into the water in the Styrofoam calorimeter cup. Stir and measure the temperature of the mixture until the temperature stabilizes. Record this stabilized temperature and the final temperature for the water (T_{fw}) and the final temperature for the metal sample (T_{fs}). Calculate the specific heat (c_s) of the metal sample. Note that ΔT_w is obtained from $|T_{fw} - T_{iw}|$ and ΔT_s is obtained from $|T_{fs} - T_{is}|$.

8. Repeat the above steps for sample 2, recording all measurement data in Data Table 18.2. Repeat the procedure for sample 3, recording all measurement data in Data Table 18.3. Run a second trial on all three samples, comparing the results of both trials on each sample. Compare the calculations from the two trials on each sample to decide if a third trial is needed.

Results

1. Calculate the specific heat (c_s) for each sample. Show all work and record your result in each data table.
2. Using the accepted value for each sample, calculate the percentage error and record it in each data table.
3. Discuss and evaluate the magnitude of various sources of error in this experiment.

4. What would happen to the calculated specific heat if some boiling water were to slosh into the cup with the metal?

5. Was the purpose of this lab accomplished? Why or why not? (Your answer to this question should show thoughtful analysis and careful, thorough thinking.)

Invitation to Inquiry

Predict what will happen if you heat a brass, glass, and iron ball to 100°C and place them on a sheet of paraffin. Test your prediction, then explain how this could happen.

Data Table 18.1   Specific Heat of _____

	Trial 1	Trial 2
Mass of sample (m_s)	_____	_____
Initial temperature of cold water (T_{iw})	_____	_____
Mass of cold water (m_w)	_____	_____
Initial temperature of metal sample (T_{is})	_____	_____
Final temperature of metal sample (T_{fs})	_____	_____
Final temperature of water (T_{fw})	_____	_____
Calculated specific heat (c_s)	_____	_____
Accepted value	_____	_____
Percent error	_____	_____

Data Table 18.2   Specific Heat of _____

	Trial 1	Trial 2
Mass of sample (m_s)		
Initial temperature of cold water (T_{iw})		
Mass of cold water (m_w)		
Initial temperature of metal sample (T_{is})		
Final temperature of metal sample (T_{fs})		
Final temperature of water (T_{fw})		
Calculated specific heat (c_s)		
Accepted value		
Percent error		

Data Table 18.3   Specific Heat of _____

	Trial 1	Trial 2
Mass of sample (m_s)	_____	_____
Initial temperature of cold water (T_{iw})	_____	_____
Mass of cold water (m_w)	_____	_____
Initial temperature of metal sample (T_{is})	_____	_____
Final temperature of metal sample (T_{fs})	_____	_____
Final temperature of water (T_{fw})	_____	_____
Calculated specific heat (c_s)	_____	_____
Accepted value	_____	_____
Percent error	_____	_____

Experiment 19: Static Electricity

Introduction

Charges of static electricity are produced when two dissimilar materials are rubbed together. Often the charges are small or leak away rapidly, especially in humid air, but they can lead to annoying electrical shocks when the air is dry. The charge is produced because electrons are moved by friction and this can result in a material acquiring an excess of electrons and becoming a negatively charged body. The material losing electrons now has a deficiency of electrons and is a positively charged body. All electric static charges result from such gains or losses of electrons. Once charged by friction, objects soon return to the neutral state by the movement of electrons. This happens more quickly in humid air because water vapor assists with the movement of electrons from charged objects. In this experiment you will study the behavior of static electricity, hopefully on a day of low humidity.

Procedure
Part A: Attraction and Repulsion

1. Rub a glass rod briskly for several minutes with a piece of nylon or silk. Suspend the rod from a thread tied to a wooden meterstick as shown in figure 19.1. Rub a second glass rod briskly for several minutes with nylon or silk. Bring it near the suspended rod and record your observations in Data Table 19.1. (If nothing is observed to happen, repeat the procedure and rub both rods briskly for twice the time.)

2. Repeat the procedure with a hard rubber rod that has been briskly rubbed with wool or fur. Bring a second hard rubber rod that has also been rubbed with wool or fur near the suspended rubber rod. Record your observations as in procedure step 1.

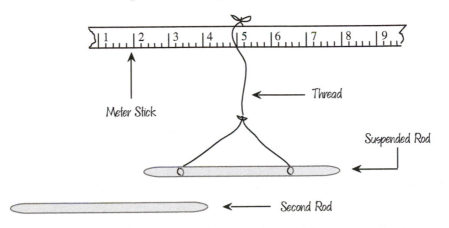

Figure 19.1

3. Again rub the hard rubber rod briskly with wool or fur and suspend it. This time briskly rub a glass rod with nylon or silk and bring the glass rod near the suspended rubber rod. Record your observations.

4. Briskly rub a glass rod with nylon or silk and bring it near, but not touching, the terminal of an electroscope (figure 19.2). Record your observations.

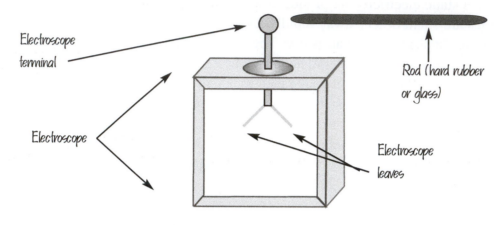

Electroscope terminal

Rod (hard rubber or glass)

Electroscope

Electroscope leaves

Figure 19.2

5. Repeat procedure step 4 with a hard rubber rod rubbed with wool or fur, again not touching the electroscope terminal. Record your observations.

Part B: Charging by Induction

1. Inflate two rubber balloons and tie the ends. Attach threads to each balloon and hang them next to each other from a support. Rub both balloons with fur or wool and allow them to hang freely. Record your observations in Data Table 19.2.

2. Bring a hard rubber rod that has been rubbed with wool or fur near the rubbed balloons. Record your observations.

3. Bring a glass rod that has been rubbed with nylon or silk near the rubbed balloons. Record your observations.

4. Detach one of the balloons by breaking or cutting the thread. Rub the balloon with fur or wool for several minutes. Hold the balloon against a wall and slowly release it. Record your observations.

5. Move the rubbed balloon near an electroscope and record your observations.

6. Move an electroscope near the wall where the balloon was held. Record your observations.

Part C: Determining the Sign of a Charge

1. When a rubbed hard rubber rod is brought near the terminal of an electroscope the leaves will stand apart but fall back together when the rod is removed.

2. When a rubbed hard rubber rod touches the terminal of an electroscope the leaves stand apart as before. When the rod is removed this time the leaves *remain* apart.

3. When the charged rod was brought near the terminal a charge was *induced* by the reorientation of charges in the terminal and leaves. When the rod was removed, the charges returned to their original orientation and the leaves collapsed because no net charge remained on the electroscope.

4. When the electrode was touched, charge was transferred to (or from) the electroscope and removing the rod had no effect on removing the charge. Touching the terminal with your finger returns the electroscope to a neutral condition.

5. An electroscope may be used to determine the sign of a charged object. First, charge the electroscope by induction as in procedure step 3 above. While the charged rod is near the terminal, touch the opposite side of the terminal with a finger of your free hand. Electrons will be repelled and conducted away through your finger. Remove your finger from the terminal, then move the rubber rod from near the electroscope. The electroscope leaves now have a net positive charge. If a charged object is brought near the electroscope the leaves will spread farther apart if the object has a positive charge. If the charged object has a negative charge, electrons are repelled into the leaves and they will move together as they are neutralized.

6. The process of an object gaining an excess of electrons or losing electrons through friction is complicated and not fully understood theoretically. It is possible experimentally, however, to make a list of materials according to their ability to lose or gain electrons. Gather various materials such as polyethylene film, rubber, wood, cotton, silk, nylon, fur, wool, glass, and plastic. Give an electroscope a positive charge by induction as described in procedure step 5. Rub combinations of the materials together and determine if the charge on each material is positive or negative. Record your findings.

Results

1. Describe two different ways that electrical charge can be produced by friction.

2. Describe how you can determine the sign of a charged object. What assumption must be made using this procedure?

3. Move a hard rubber rod that has been rubbed with wool or fur near a very thin, steady stream of water from a faucet. Describe, then explain your observations.

4. Was the purpose of this lab accomplished? Why or why not? (Your answer to this question should be reasonable and make sense, showing thoughtful analysis and careful, thorough thinking.)

Invitation to Inquiry

Make a charge detector. Spray two grains of puffed rice or wheat with a very thin layer of aluminum paint. Use a needle and thread to suspend the grains from a stopper or cork. Use this charge detector to investigate charged plastic rods, combs, glass rods, and other items.

Data Table 19.1	Attraction and Repulsion of Glass Rod and Rubber Rod

Interaction	Observations
Glass rod - Glass rod	
Rubber rod - Rubber rod	
Glass rod - Rubber rod	
Glass rod - Electroscope	
Rubber rod - Electroscope	

How many kinds of electric charge exist according to your findings above? Explain your reasoning.

How do charges interact?

Data Table 19.2   Charging by Induction

Interaction	Observations
Balloon - Balloon	
Rubber rod - Balloon	
Glass rod - Balloon	
Balloon - Wall	
Balloon - Electroscope	
Wall - Electroscope	

What evidence did you find to indicate that the balloons had static charges:

Describe the evidence you found to indicate that the wall was or was not charged as shown by the electroscope. Explain.

Explain why a balloon exhibits the behavior that it did on the wall.

Experiment 20: Electric Circuits

Introduction

An electric *circuit* is a conducting path for an electric *current*, which is a flow of charge. When the circuit is connected to a battery, for example, charges flows from one terminal of the battery, through one or more electrical devices, and then back to the other terminal. Circuits can be described with words or by symbols that are widely known and used. Some of these symbols are illustrated in figure 20.1.

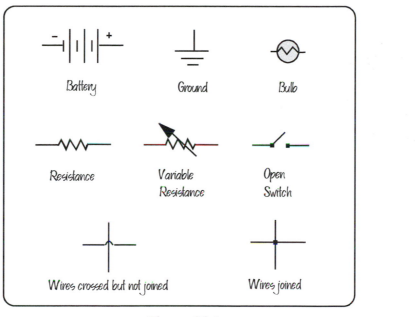

Figure 20.1

Procedure

1. Use *one* flashlight battery, *one* flashlight bulb, and *one* 10 cm piece of hook-up wire to make the bulb light. On page 180, make a sketch of your circuit. Use the symbols given in the introduction to show exactly how you constructed the circuit to make the bulb light. Draw an arrow to show the flow of **conventional current** and a second arrow to show the flow of the **electron current**. Identify both arrows.

2. Note the construction of the flashlight bulb. Figure 20.2 is an incomplete diagram of a bulb. Complete this diagram by drawing wires that connect the filament to the contact points in such a way that there is a complete circuit for the current to follow.

3. Wire the circuits that are shown in figure 20.3. For each circuit, note if a single bulb lights dimly, brightly, or if at all, and note when two bulbs are lit if unscrewing one affects the other in any way. Record these and other observations in Data Table 20.1.

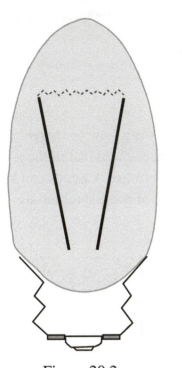

Figure 20.2

Sketch of your circuit of one battery, one bulb, and one wire.

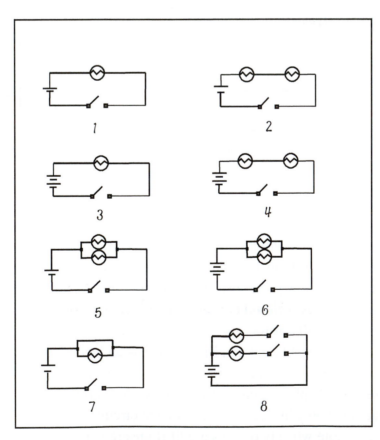

Figure 20.3

Results

1. Using the brightness of the bulb in the simple circuit (#1) as the standard for normal, describe which circuits had bulbs with a normal brightness. (In your attempt to control variables, note the color of the small glass blob at the base of the filament in each bulb. Different colors indicate bulbs of different resistance.)

2. Which circuits had bulbs that were dimmer than normal? Explain.

3. Which circuits had bulbs that were brighter than normal? Explain.

4. In which circuits did removing one bulb cause the other to go out? Explain.

5. In which circuits did removing one bulb not affect the other? Explain.

6. Describe the general requirements for any circuit to function.

7. Was the purpose of this lab accomplished? Why or why not? (Your answer to this question should show thoughtful analysis and careful, thorough thinking.)

Invitation to Inquiry

Obtain a lemon and roll it a few times on the table top. Make two parallel slits very close together in the lemon and insert a clean strip of magnesium in one slit and a clean strip of copper in the other slit. The strips can be very close but must not touch each other. Use alligator clips and see if you can run a small motor or bulb with this cell. Try the two metals in orange juice, a potato, apples, soft drinks, and other substances. Try two different metals. How long can you run a clock or some other device? Which substance is best for running a clock or bulb? Which two metals?

Data Table 20.1	Circuit and Bulb Brightness Relationships
Circuit	Observations
Circuit 1, a simple circuit.	
Circuit 2, two bulbs in series.	
Circuit 3, two batteries in series.	
Circuit 4, two bulbs in series with two batteries in series.	
Circuit 5, two bulbs in parallel.	
Circuit 6, two bulbs in parallel and two batteries in series.	
Circuit 7, one bulb in parallel with wire. Show with arrows how current takes path of least resistance.	
Circuit 8, switches in series with bulbs in parallel circuit.	

Experiment 21: Series and Parallel Circuits

Introduction

There are two basic ways to connect more than one resistance in a circuit, in a *series circuit* or in a *parallel circuit*. A **series circuit** has each resistance connected one after the other so the same charges flows through one resistance, then the next one, and so on. A **parallel circuit** has separate pathways for the charges so they do not go through one resistance after the other. The use of the term "parallel" means that charges can flow through one of the branches, but not the others. The term does not mean that the branches are necessarily lined up with each other. The series and parallel circuits have separate characteristics that offer certain advantages and disadvantages.

Adding more resistances in a series circuit results in two major effects that are characteristic of all series circuits. More resistances result in (1) a *decrease in the current* available in the circuit, and (2) a *reduction of the voltage* available for each resistance. Since power is determined from $I \times V$, adding more lamps will result in dimmer lights. Perhaps you have observed such a dimming when you connected two strings of Christmas tree lights. Many Christmas tree lights are connected in a series circuit. Another disadvantage to a series circuit is that if one bulb burns out the circuit is broken and all the lights go out.

Adding more resistances in a parallel circuit results in three major effects that are characteristic of all parallel circuits. More resistances result in (1) an *increase in the current* in the circuit, (2) as *the same voltage* is maintained across each resistance, and (3) *lowering of the total resistance* of the entire circuit. The total resistance is lowered since additional branches provide more pathways for the charges to move.

In this investigation you will investigate some of the characteristics of series and parallel circuits as you measure the potential difference, current, resistance, and calculate the power of the same bulb in both circuits.

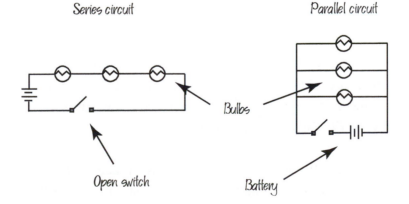

Series circuit   Parallel circuit

Bulbs

Open switch   Battery

Figure 21.1

Procedure

You will be using three #41 bulbs (0.5 A), bulb sockets, two dry cells (or other source of 3 V dc), a dc voltmeter (0 to 5 V range), and a dc ammeter (0 to 2 A range) to read potential difference and current values at various points in a series circuit and in a parallel circuit. Note that an ammeter is always connected in series in a circuit so all the desired current passes through it. A voltmeter is always wired in parallel so that it measures the difference in potential between two points in the circuit. Also note that one of the terminals of the dc meter is marked negative (–). Always connect this negative (–) terminal of the meter to the negative (–) terminal of a battery (or other voltage source). If you connect the (–) of the meter to the (+) of a voltage source the pointer of the meter will move backward off the dial, which could damage the meter.

1. Connect two dry cells in series, then wire the connected cells to three lamps and a switch in series as shown in figure 21.2.

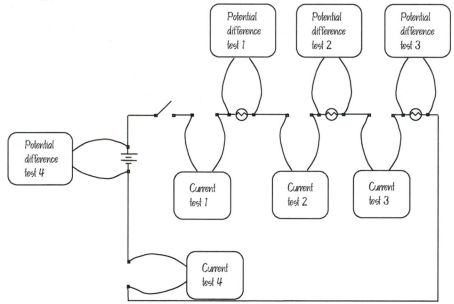

Figure 21.2

2. Make four separate voltage readings with a voltmeter at the places in the circuit as shown in figure 21.2. Note that test 1 measures the voltage drop for bulb 1, test 2 measures the voltage drop for bulb 2, test 3 measures the voltage drop for bulb 3, and test 4 measures the voltage source for the entire circuit. Record your findings in Data Table 21.1.

3. Make four separate current readings with an ammeter at the places in the circuit as shown in figure 21.2. Note that test 1 measures the current before bulb 1, test 2 measures the current before bulb 3, test 3 measures the current before bulb 3, and test 4 measures the current for the entire circuit. Record your findings in Data Table 21.1.

4. Remove one of the bulbs from the circuit and record here the effect on the other lamps:

5. Connect the three lamps in parallel as shown in figure 21.3.

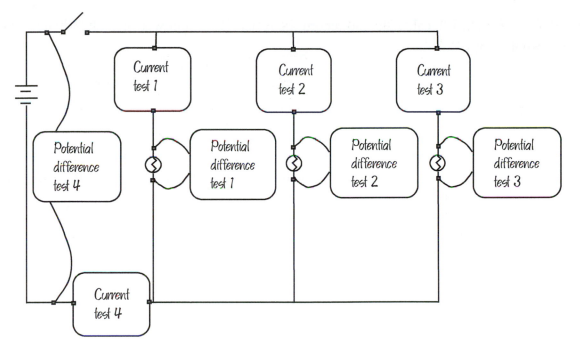

Figure 21.3

6. Make four separate voltage readings with a voltmeter at the places in the circuit as shown in figure 21.3. Note that test 1 measures the voltage drop for bulb 1, test 2 measures the voltage drop for bulb 2, test 3 measures the voltage drop for bulb 3, and test 4 measures the voltage source for the entire circuit. Record your findings in Data Table 21.2.

7. Make four separate current readings with an ammeter at the places in the circuit as shown in figure 21.3. Note that test 1 measures the current before bulb 1, test 2 measures the current before bulb 2, test 3 measures the current before bulb 3, and test 4 measures the current for the entire circuit. Record your findings in Data Table 21.2.

8. Remove one of the lamps from the circuit and record here the effect on the other lamps:

9. In Data Table 21.1, calculate and record the resistance and power of each bulb and for the entire series circuit.

10. In Data Table 21.2, calculate and record the resistance and power of each bulb and for the entire parallel circuit.

Results

1. How does the voltage drop of each of the three bulbs in the series circuit compare with the voltage of the source?

2. How does the voltage drop of each of the three bulbs in the parallel circuit compare with the voltage of the source?

3. How does the current through one bulb in the series circuit compare with the total current through the three lamps?

4. How does the current through one bulb in the parallel circuit compare with the total current through the three lamps?

5. Account for any differences in brightness observed by the same bulbs in series and parallel circuits.

6. According to the results of this investigation, what happens to the current and voltage available for each resistance as more are added (a) to a series circuit, (b) to a parallel circuit?

7. Other than differences in current, voltage, and resistance, what distinguishing characteristic will tell you if a circuit is a series or parallel circuit?

8. Was the purpose of this lab accomplished? Why or why not? (Your answer to this question should show thoughtful analysis and careful, thorough thinking.)

Invitation to Inquiry

Make a model of a way to wire a circuit in two rooms. Arrange bulbs and switches so a person walking into one side of a room can turn on the lights, then turn on the lights in a second room so the lights in the first room turn off at the same time.

Bulb or Circuit	Volts (V)	Amps (A)	Resistance (Ω)	Power (W)
Data Table 21.1	Series Circuit			
1	_____	_____	_____	_____
2	_____	_____	_____	_____
3	_____	_____	_____	_____
Whole Circuit	_____	_____	_____	_____

Data Table 21.2	Parallel Circuit			
Bulb or Circuit	Volts (V)	Amps (A)	Resistance (Ω)	Power (W)
1	_____	_____	_____	_____
2	_____	_____	_____	_____
3	_____	_____	_____	_____
Whole Circuit	_____	_____	_____	_____

Experiment 22: Ohm's Law

Introduction

An electric charge has an electric field surrounding it, and work must be done to move a like-charged particle into this field since like charges repel. The electrical potential energy is changed just as gravitational potential energy is changed by moving a mass in the earth's gravitational field. A charged particle moved into the field of a like-charged particle has potential energy in the same way that a compressed spring has potential energy. In electrical matters the potential difference that is created by doing work to move a certain charge creates electrical potential. A measure of the electrical potential difference between two points is the **volt** (V).

A volt measure describes the potential difference between two places in an electric circuit. By analogy to pressure on water in a circuit of water pipes the potential difference is sometimes called an "electrical force" (emf). Also by analogy to water in a circuit of water pipes, there is a varying rate of flow at various pressures. An electric **current** *(I)* is the quantity of charge moving through a conductor in a unit of time. The unit defined for measuring this rate is the **ampere** (A), or the **amp** for short.

The rate of water flow in a pipe is directly proportional to the water pressure; e.g., a greater pressure produces a greater flow. In an electric circuit the current is directly proportional to the potential difference (V) between two points. Most materials, however, have a property of opposing or reducing a current, and this property is called **electrical resistance** (R). If a conductor offers a small resistance less voltage would be required to push an amp of current through the circuit. On the other hand, a greater resistance requires more voltage to push the same amp of current through the circuit. Resistance (R) is therefore a *ratio* of the potential difference (V) between two points and the resulting current. This ratio is the unit of resistance and is called an **ohm** (Ω). Another way to show the relationship between the voltage, current, and resistance is

$$R = \frac{V}{I}$$

or

$$V = IR$$

which is known as **Ohm's law**. This is one of the three ways to show the relationship; this one (solved for V) happens to be the equation of a straight line with a slope *R* when V is on the *y*-axis, *I* is on the *x*-axis, and the *y*-intercept is zero.

Procedure

Part A: Known Resistance

1. A known resistance will be provided for use in this circuit:

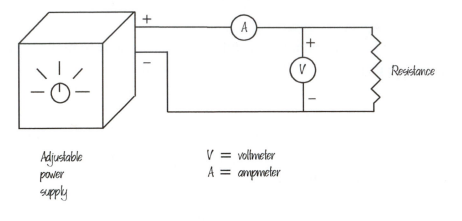

Figure 22.1

2. You will adjust the dc adjustable power supply as instructed by your laboratory instructor, obtaining six values for voltage and current using the supplied resistor. Set up the circuit with the power *off* and do not proceed until the laboratory instructor has checked the circuit.

3. Record the value of the resistor and the six values for the current and voltage in Data Table 22.1.

Part B: Unknown Resistance

Repeat procedure A with an unknown resistor. Record your data in Data Table 22.2.

Results

1. Make a graph of the six data points of Data Table 22.1, placing the current on the *x*-axis and the voltage on the *y*-axis. Calculate the slope and write it here and somewhere on the graph.

2. Compare the calculated value of the known resistor with the accepted value as given by your instructor. Calculate the percentage error.

3. Make a second graph, this time of the six data points in Data Table 22.2, again placing the current on the *x*-axis and the voltage on the *y*-axis. Calculate the slope and write it here and somewhere on the graph.

4. What is the value of the unknown resistor?

5. Explain how the two graphs demonstrate Ohm's law.

6. Was the purpose of this lab accomplished? Why or why not? (Your answer to this question should show thoughtful analysis and careful, thorough thinking.)

Going Further

1. Check your answer about the value of the unknown resistor by using your calculated value in the equation of a straight line when V = 2 V, 4 V, and 6 V. Verify with the laboratory equipment and calculate the average percentage error. Describe your results here:

2. Use three different resistances (e.g., 16 Ω, 30 Ω, and 47 Ω) connected in a series for four different input voltages (2 V, 4 V, 6 V, and 8 V) and connected in a parallel circuit. Plot voltage vs. total current for both the series and parallel circuits and quantitatively show how the total resistance (the slope) differs for series and parallel circuits.

Invitation to Inquiry

The carbon resistors that are used as standard sources of resistance in electrical circuits are marked with a code of colored bands. Here is the code for the colors:

Black = 0	Green = 5
Brown = 1	Blue = 6
Red = 2	Violet = 7
Orange = 3	Gray = 8
Yellow = 4	White = 9

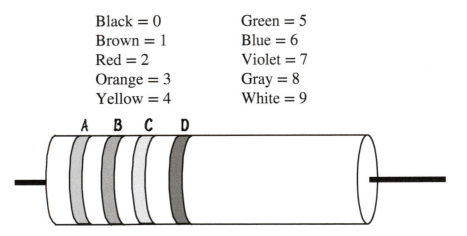

The value of the resistor is AB × 10^C ±D where no D band means ±20%, silver means ±10%, and gold means ±5%. The band placement is shown above. As an example, consider bands of red, yellow, red, and silver on a resistor. This means 24 × 10^2 ±10% ohms, or 2400 ±240 Ω.

Obtain 5 or 6 resistors and a meter to measure the experimental resistance of each. Read the code to determine the accepted value, then find the experimental error as described in the Appendix. What could account for experimental errors, if any?

Data Table 22.1	Voltage and Current Relationships With Known Resistance	
Trial	Voltage (V)	Current (A)
1	_____	_____
2	_____	_____
3	_____	_____
4	_____	_____
5	_____	_____
6	_____	_____

Resistor _____ Ω

Data Table 22.2	Voltage and Current Relationships With Unknown Resistance	
Trial	Voltage (V)	Current (A)
1	_____	_____
2	_____	_____
3	_____	_____
4	_____	_____
5	_____	_____
6	_____	_____

Resistor _____Ω

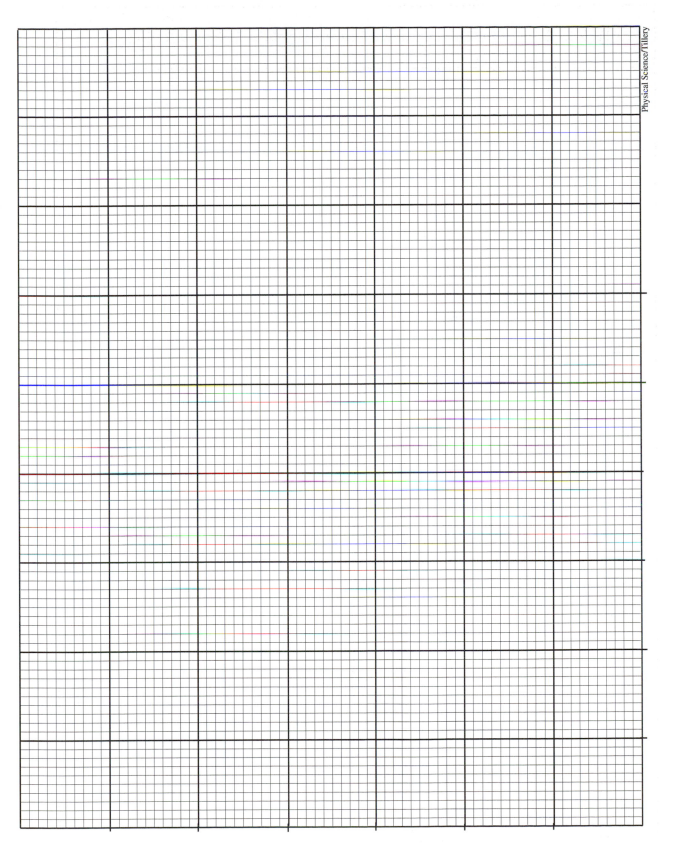

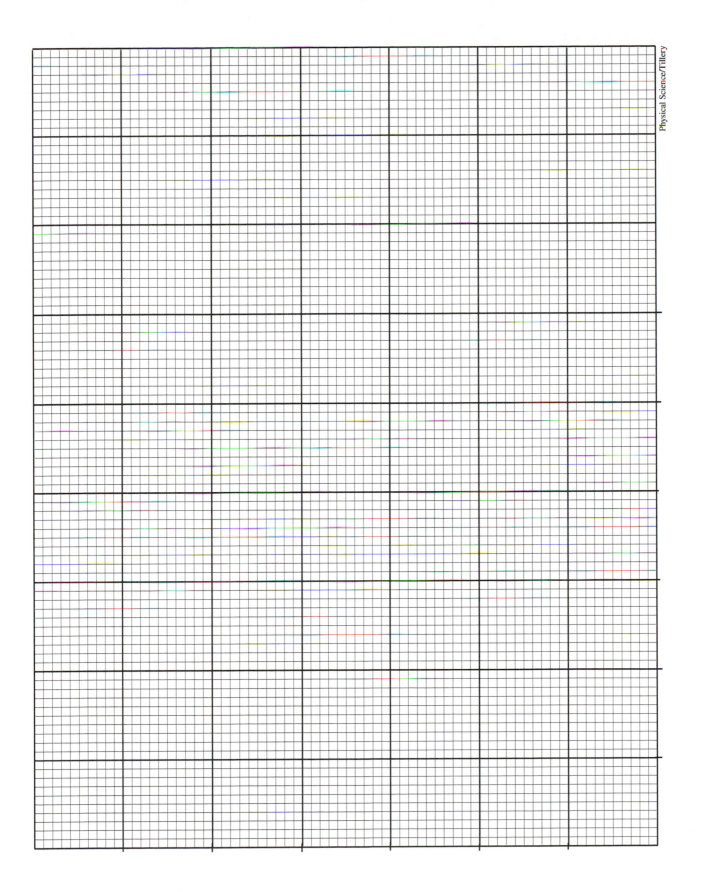

Experiment 23: Magnetic Fields

Introduction

A magnet moved into the space near a second magnet experiences a force as it enters the **magnetic field** of the second magnet. The magnetic field model is a conceptual way of considering how two magnets interact with one another. The magnetic field model does not consider the force that one magnet exerts on another one through a distance. Instead, it considers *the condition of space* around a magnet. The condition of space around a magnet is considered to be changed by the presence of the magnet. Since this region of space, or field, is produced by a magnet, it is called a *magnetic field*. A magnetic field can be represented by *magnetic field lines*. By convention, magnetic field lines are drawn to indicate how the *north pole* of a tiny imaginary magnet would point when in various places in the magnetic field. Arrowheads indicate the direction that the north pole would point, thus defining the direction of the magnetic field. The strength of the magnetic field is greater where the lines are closer together and weaker where they are further apart. Magnetic field lines emerge from a magnet at the north pole and enter the magnet at the south pole. Magnetic field lines always form closed loops.

Magnetic field strength is defined in terms of the magnetic force exerted on a test charge of a particular charge and velocity. The magnetic field is thus represented by vectors (symbol B) which define the field strength, also called the magnetic induction. The units are:

$$ B = \frac{\text{newton}}{(\text{coulomb})\left(\dfrac{\text{meters}}{\text{second}}\right)} $$

Since a coulomb/s is an amp, this can be written as

$$ B = \frac{\text{newton}}{\text{amp} \cdot \text{meter}} $$

which is called a **tesla** (T). The tesla is a measure of the strength of a magnetic field. Near the surface, the earth's horizontal magnetic field in some locations is about 2×10^{-5} tesla. A small bar magnet produces a magnetic field of about 10^{-2} tesla but large, strong magnets can produce magnetic fields of 2 tesla. Superconducting magnets have magnetic fields as high as 30 tesla. Another measure of magnetic field strength is called the **gauss** (G) (1 tesla $= 10^4$ gauss). Thus the process of demagnetizing something is sometimes referred to as "degaussing."

In this experiment you will investigate the magnetic field around a permanent magnet and the magnetic field around a current-carrying conductor.

Procedure

1. Tape a large sheet of paper on a table, with the long edge parallel to the north-south magnetic direction as determined by a compass.

2. Center a bar magnet on the paper with its south pole pointing north. Use a sharp pencil to outline lightly the bar magnet on the paper. Write N and S on the paper to record the north-seeking and south-seeking poles of the bar magnet. Place the bar magnet back on its outline if you moved it to write the N and the S.

3. Slide a small magnetic compass across the paper, stopping close to the north-seeking pole of the bar magnet. Make two dots on the paper, one on either side of the compass and aligned with the compass needle. See figure 23.1.

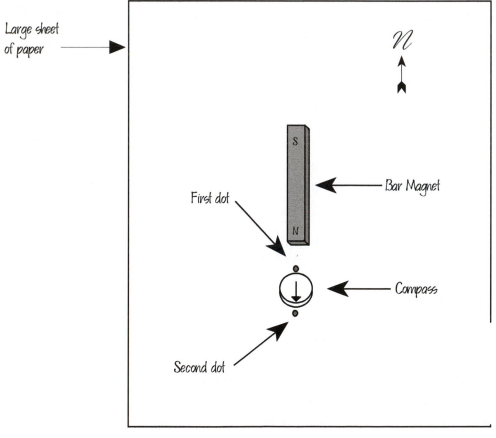

Figure 23.1

4. Slide the compass so the south pole of the needle is now directly over the dot that was at the north pole of the needle. Make a new dot at the north pole end of the compass, exactly in front of the needle. See figure 23.2.

5. Continuing the process of moving the compass so the south pole of the needle is over the most recently-drawn dot, then making another new dot at the north pole of the needle. Stop when you reach the bar magnet or the edge of the paper.

6. Draw a smooth curve through the dots, using several arrowheads to show the direction of the magnetic flux line.

7. Repeat procedure steps 3 through 6, by starting with the compass in a new location somewhere around the bar magnet. Repeat the procedures until enough flux lines are drawn to make a map of the magnetic field.

8. Place a second large sheet of paper on a large rigid plastic sheet (or glass plate) on top of the bar magnet. Sprinkle a thin, even layer of iron filings over the plastic, tapping the sheet lightly as you sprinkle. Sketch the magnet flux lines on the paper as shown by the arrangement of the filings.

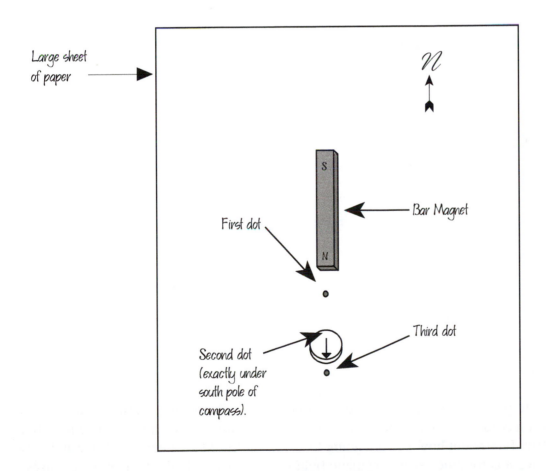

Figure 23.2

Results

1. How is using iron filings (a) similar, and (b) different from using a magnetic compass to map a magnetic field?

2. In terms of a force, or torque on a magnetic compass needle, what do the lines actually represent? Explain.

3. Do the lines ever cross each other at any point? Explain.

4. Where do the lines appear to be concentrated the most? What does this mean?

Invitation to Inquiry

Is it possible to produce electricity in an extension cord that is not plugged into a circuit? Hook a 50 ft extension cord to a galvanometer and move it as a jump rope, cutting the magnetic field lines around the earth. Figure out how you are going to attach the cord to the galvanometer and how you are going to move it across the earth's magnetic field lines. Can you think of any practical uses for "jump-rope electricity?"

Experiment 24: Electromagnets

Introduction

Electric charges in motion produce a magnetic field around the charges, and a current carrying wire has a magnetic field as a result. You can "map" such a magnetic field by running a straight wire vertically through a sheet of paper. The wire is connected to a battery and iron filings are sprinkled on the paper. The filings will become aligned as the domains in each tiny piece of iron are forced parallel to the field. Overall, filings near the wire form a pattern of concentric circles with the wire in the center.

The direction of the magnetic field around a current-carrying wire can be determined by using the common device for finding the direction of a magnetic field, the magnetic compass. The north-seeking pole of the compass needle will point in the direction of the magnetic field lines (by definition). If you move the compass around the wire the needle will always move to a position that is tangent to a circle around the wire. Evidently the magnetic field lines are closed, concentric circles that are at right angles to the length of the wire. If you *reverse* the direction of the current in the wire and again move a compass around the wire, the needle will again move to a position that is tangent to a circle around the wire. This time the north pole direction is reversed. Thus the magnetic field around a current-carrying wire has closed concentric field lines that are perpendicular to the length of the wire. The *direction* of the magnetic field is determined by the direction of the current.

The *strength* of the magnetic field (*B*) around a long, straight current-carrying wire is directly proportional to the current (*I*) in the wire and inversely proportional to the distance (*d*) from the wire, or

$$B = k\frac{2I}{d}$$

where the proportionality constant k is 1.00×10^{-7} newton/amp^2. Note that the magnetic field strength varies with the distance from the wire (not the square of the distance) and that the unit is newton/amp·meter, or tesla.

A current-carrying wire that is formed into a loop has perpendicular, circular field lines that pass through the inside of the loop in the same direction. This has the effect of concentrating the field lines which increases the magnetic field intensity. Since the field lines all pass through the loop in the same direction, one side of the loop will have a north pole and the other side a south pole. Many loops of wire formed into a cylindrical coil is called a **solenoid**. When a current is in a solenoid, each loop contributes field lines along the length of the cylinder. The overall effect is a magnetic field around the solenoid that acts just like the magnetic field of a bar magnet. This magnet, called an **electromagnet,** can be turned on or off by turning the current on and off. In addition, the strength of the electromagnet depends on the magnitude of the current and the number of loops (ampere-turns). The strength of the electromagnet can also be increased by placing a piece of soft iron in the coil. The magnetic domains of the iron become aligned by the influence of the magnetic field. This induced magnetism increases the overall magnetic field strength of the solenoid as the magnetic field lines are gathered into a smaller volume within the core.

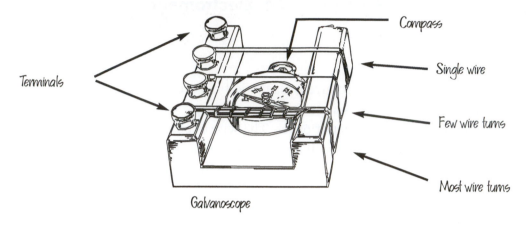

Figure 24.1

Procedure

1. Place a galvanoscope on a nonmetallic table with the wire coils running in a north-south direction. Place a compass beneath the coil of wires with the most turns. Situate the compass and the device so the needle is parallel and beneath the wire coil, with the north pole of the compass needle pointing to the zero degree mark.

2. Connect a dry cell or laboratory source of current at 1.5 V to the terminal of most turns. Note the direction of the current (north-to-south or south-to-north), the direction of the needle deflection (east or west), and the amount of deflection in degrees. Disconnect the source of current. Record your observations in Data Table 24.1.

3. Reverse the direction of the current and repeat the observations and recordings of procedure step 2. Disconnect the source of current, then record your observations in Data Table 24.1.

4. With the current disconnected, carefully move the compass to beneath the coil of wires with a few turns, otherwise situating the compass and device exactly as they were in procedure step 1. Repeat procedure steps 2 and 3 with the compass beneath this coil and record your observations in Data Table 24.1.

5. Repeat procedure step 4, this time moving the compass to beneath the single wire. Record all observations as before.

6. Obtain about 3 m of No. 18 insulated copper wire and a 1/2 cm diameter soft iron spike that is about 12 cm long. This is sufficient wire to leave about 20 cm free, then wrap about 100 turns around the spike, leaving about 20 cm free at the opposite end. Use this device and a dry cell or

laboratory source of current at 1.5 V to begin a series of experiments to find out what factors affect the strength and polarity of an electromagnet. The strength of the electromagnet could be measured in terms of how many paper clips or nails it will pick up. Be sure to record all procedures tried as well as the results, remembering that "no result" is a finding as well as more dramatic events.

Results

1. What determines the *direction* of a magnetic field around a current-carrying wire? Provide evidence for your answer.

2. What determines the *strength* of a magnetic field around a current-carrying wire? Provide evidence for your answer.

3. Which is stronger, an electromagnet with an iron core or an electromagnet without an iron core? Explain.

4. Was the purpose of this lab accomplished? Why or why not? (Your answer to this question should show thoughtful analysis and careful, thorough thinking.)

Invitation to Inquiry

Will one electromagnet attract or repel another electromagnet when there is a current in both coils? Test this idea with an experiment after recording your prediction.

Data Table 24.1	Direction and Strength of Magnetic Field Around Electromagnet		
Wire Coil	Direction of Current	Direction of Deflection	Amount of Deflection
Most turns	_____	_____	_____
Most turns (reverse current)	_____	_____	_____
Several turns	_____	_____	_____
Several turns (reverse current)	_____	_____	_____
Single wire	_____	_____	_____
Single wire (reverse current)	_____	_____	_____

Experiment 25: Standing Waves

Introduction

A stringed musical instrument, such as a guitar, violin, or piano, has strings that are stretched between two fixed ends. When a string is plucked, bowed, hit, or otherwise disturbed, waves of many different frequencies will travel back and forth on the string, reflecting from the fixed ends. Many of these waves quickly fade away but certain frequencies **resonate**, setting up patterns of waves. Reflected waves interfere with incoming waves of the same frequency, making (1) stationary places of destructive interference called **nodes**, which show no disturbance, and (2) loops of constructive interference called **antinodes**. Antinodes form where the crests and troughs of the two wave patterns produce a disturbance that rapidly alternates upward and downward. The pattern of alternating nodes and antinodes does not move along the string and is thus called a **standing wave.** A standing wave *for one wavelength* will have three nodes, one at each end and one in the center, and will have two antinodes. Standing waves occur at the natural, or resonant, frequency of a string and are determined by the length, tension, and density (mass per unit length) of the string.

Standing waves are produced when a condition of resonance exists between the natural frequency of a string and the frequency of a disturbance. At resonance, there is a particular wavelength (λ) that is directly proportional to the velocity of the wave along the string and is inversely proportional to the frequency, or $v = f\lambda$, where f is the frequency and v is the velocity. This velocity is determined by the tension in the string (F_T) and the density (D) or mass per unit length of the string. The velocity is related to the tension (F_T) and the density (D) by

$$v = \sqrt{\frac{F_T}{D}}$$

Changing the tension thus changes the velocity. Changes in the velocity result in changes in the wavelength at a constant frequency since $v = f\lambda$. Therefore, changing the tension at a constant frequency will result in different numbers of standing waves as the conditions for resonance are met by a varying tension.

In this experiment standing waves are set up in a stretched nylon string held under tension by masses ($F_T = mg$) at one end and a vibrator at the other end. The vibrator is connected to a 60-cycle alternating current, which drives the vibrator at the power-line frequency (60 Hz) or else twice that frequency (120 Hz) depending on the orientation of the vibrator relative to the string. Most of the time the frequency will be double the supply current, or 120 Hz. A stroboscope can be used to check the frequency if there is a question.

The tension in the string (F_T) is measured from the masses ($F_T = mg$) suspended over a pulley with a weight hanger and is changed by adding or removing masses.

The length of a wave (λ) is twice the distance (L) between two consecutive nodes in a standing wave, so $\lambda = 2L$.

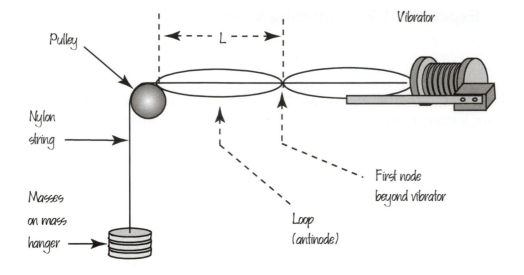

Figure 25.1

The relationship between the velocity (v), wavelength (λ), and frequency (f) is found in the wave equation, $v = f\lambda$. Solving for wavelength gives

$$\lambda = \frac{v}{f}$$

But the velocity is related to the tension in the string (F_T) and the density (D) (i.e., mass per unit length) by

$$v = \sqrt{\frac{F_T}{D}}$$

Substituting the tension and density variables for velocity in the wave equation gives a wavelength λ of

$$\lambda = \frac{\sqrt{\frac{F_T}{D}}}{f}$$

or

$$\lambda = \left(\frac{1}{f}\right)\left(\sqrt{\frac{F_T}{D}}\right)$$

At a constant frequency of vibration (f) and string density (D), the wavelength (λ) and the square root of the tension should therefore be proportional, with a proportionality constant of

$$\text{Slope} = \frac{1}{\left(f\sqrt{D}\right)}$$

212

Procedure

1. Obtain a length of nylon string of a known density, or measure a 150-cm length and determine the mass per unit length with a balance. In either case, record the density (D) or mass per unit length in Data Table 25.1 in kg/m. Use one string only in this procedure.

2. Fasten one end of the string to the vibrator and pass the other end over a pulley about 1 meter away. Fasten a mass hanger to the end over the pulley (figure 25.1). Set the string length to 1.00 m.

3. Turn on the vibrator and add masses to the mass hanger until the string makes a standing wave of two definite segments. Adjust the tension by adding or removing small masses until the amplitude is at a maximum. If necessary, loosen the clamp holding the vibrator and carefully move it very slightly for maximum amplitude.

4. Measure the distance from the first node beyond the vibrator to the next node. The distance between two consecutive nodes is one-half wavelength, so two segments of measured length on the string equal one wavelength. Record the wavelength in meters and the tension in newtons for this resonant frequency.

5. Turn the vibrator on and adjust the tension by removing masses until standing waves of 3, 4, 5, 6, and 7 segments form for five separate trials. In each trial measure one segment and record the wavelength in meters with the corresponding tension in newtons.

Results

1. Plot the square root of the tension $\left(\sqrt{F_T}\right)$ in newtons $\left(\sqrt{N}\right)$ on the x-axis and the wavelength (λ) (in m) on the y-axis, using the entire graph paper. You should have a total of six data points.

2. Calculate the slope, showing your work here.

3. Determine the frequency (*f*) from the calculated slope and use this as the calculated average value for *f*.

4. Analyze the possible sources of error in this experiment.

5. What conclusions can you reach about the relationship between resonance and tension on a vibrating string?

6. What determines the quality of a musical note created by a vibrating string?

7. Was the purpose of this lab accomplished? Why or why not? (Your answer to this question should show thoughtful analysis and careful, thorough thinking.)

Invitation to Inquiry

Investigate the "Donald Duck effect" that takes place when one breathes helium. Your voice is normally produced by a stream of air flowing between the vibrating vocal chords in the larynx. The sound of your voice is also determined by the configuration of your throat, mouth, and nasal cavities. Sound waves bouncing back and forth in a cavity will interfere constructively, making standing waves known as the resonance frequencies. The cavities in the vocal tract have such resonances, and a resonance frequency will be strongly transmitted, while other frequencies will be damped.

But what happens when one breathes helium? Does your voice change to that of a high-pitched Donald Duck because helium is a low density gas? This is an invitation to investigate why the pitch of a person's voice changes. Will certain mixtures of helium and oxygen produce different effects? Warning: This is an invitation to work out the theory behind the "Donald Duck effect." Any experimental verifications must be approved ahead of time and then supervised by your instructor.

Data Table 25.1   Resonant Relationships of a String Under Tension

Number of Segments	Length of one Segment (m)	Wave Length (λ) (m)	Tension (F_T) (N)	Square Root of Tension ($\sqrt{F_T}$) ($\sqrt{N}$)
2	_____	_____	_____	_____
3	_____	_____	_____	_____
4	_____	_____	_____	_____
5	_____	_____	_____	_____
6	_____	_____	_____	_____
7	_____	_____	_____	_____

Density (D) (mass/length) _____kg/m

Frequency (vibrations/s) _____Hz

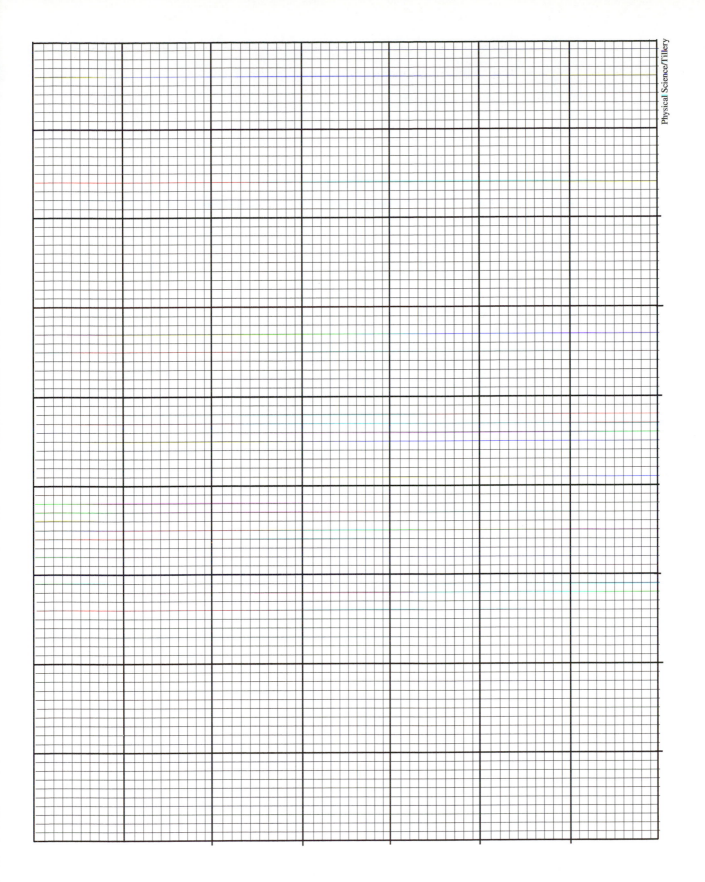

Experiment 26: Speed of Sound in Air

Introduction

A vibrating tuning fork sends a series of condensations and rarefactions through the air. When the tuning fork is held over a glass tube that is closed at the bottom, the condensations and rarefactions are reflected from the bottom. At certain lengths of tube, the reflected condensations and rarefactions are in phase with those being sent out by the tuning fork and an increase of amplitude occurs from the resonant condition. Figure 26.1 shows a wave trace representing one wavelength in which the reflected wave is in phase with the incoming wave, forming a standing wave. The antinodes represent places of maximum vibration and increased amplitude.

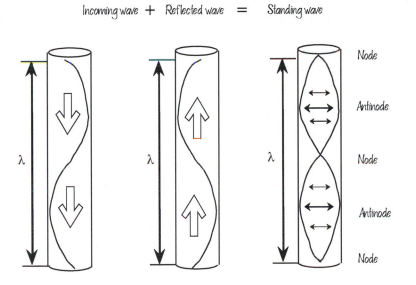

Figure 26.1

Resonance occurs when the length of the tube is such that an antinode (the place of maximum vibration) occurs at the open end. As you can see from the sketch above, there are two situations when this would occur for tube lengths less than one wavelength, 1/4 of the way up and 3/4 of the way up from the bottom. Thus resonance occurs when the length of the tube (L) is equal to 1/4 λ, 3/4 λ, 5/4 λ, and so forth where λ is the wavelength of the sound wave produced by the tuning fork.

In this experiment, a vibrating tuning fork is held just above a cylinder that is open at one end. The length to the closed end is adjusted by adding or removing water. The lowest frequency (the fundamental frequency) occurs when the longest wavelength has an antinode at the open end, so the length of the open tube is about 1/4 of the wavelength of the fundamental frequency as shown in figure 26.2. Since the length of the tube at this fundamental frequency is $L = 1/4\ \lambda$, then the fundamental wavelength must be $\lambda = 4L$.

Using the wave equation

$$v_T = f\lambda$$

and substituting the known frequency of the tuning fork for f and the experimentally determined value for the wavelength λ, you can calculate the speed of sound v_T in the tube at room temperature by using the relationship

$$v_T = v_{0°C} + \left(\frac{0.6 \text{ m/s}}{°C}\right)(T_{room})$$

where $v_{0°C}$ is the speed of sound at $0°$ C (331.4 m/s) and T_{room} is the present room temperature in $°C$.

Procedure

1. The water level in the glass tube is adjusted by raising and lowering the supply tank. Adjust the tank so the glass tube is nearly full of water.

2. Strike the tuning fork with a rubber hammer and hold the vibrating tines just above the opening of the tube.

3. Lower the water level slowly while listening for the increase in the intensity of the sound that comes with resonance. Experiment with the *entire length of the tube*, seeing how many different places of resonance you can identify.

4. Using the information learned in procedure step 3, go to the resonance level immediately *below the resonance position* of the highest water level as shown in figure 26.2. (Make sure there is *not* another resonance point between the highest water level and this second level.) Slightly raise and lower the water level until you are sure that you have found the maximum intensity. Note the relationship between the wavelength and the length of the tube as shown in figure 26.2. Measure and record in Data Table 26.1 on page 224 the length of this resonating air column to the nearest millimeter. Change the water level and run two more trials, again locating the distance with the maximum sound. Record these two lengths in Data Table 26.1 and average the length for the three trials. Record the frequency of the tuning fork (usually stamped on the handle) and the room temperature.

5. Repeat procedure steps 1 through 4 for the second resonance point *at the highest water level,* with an air column about one-third the length of the first as shown in figure 26.3. (Again, make sure there is *not* another resonance point between the highest water level and this second level.) Note the relationship between the wavelength and the length of the tube as shown in figure 26.3. Run three trials at this position and record the data in Data Table 26.2 on page 224 and, as before, average the three trials. Record the frequency of the tuning fork and the room temperature (do not assume that the room temperature remains constant).

6. Repeat the entire procedure using a different tuning fork with a different frequency. Record all data in Data Tables 26.3 and 26.4 on page 225.

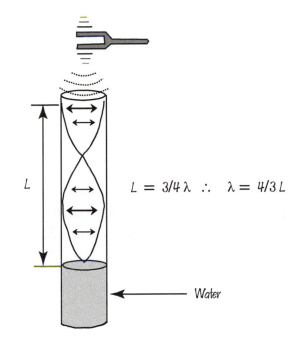

$L = 3/4 \lambda \quad \therefore \quad \lambda = 4/3 L$

Water

Figure 26.2

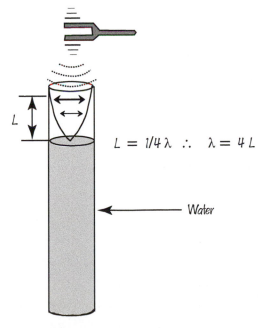

$L = 1/4 \lambda \quad \therefore \quad \lambda = 4 L$

Water

Figure 26.3

Results

1. Calculate the velocity of sound at room temperature for both tuning forks at both resonance positions and record in data tables at the measured room temperatures. Here, write the average values for both tuning forks:

2. Using the accepted value of sound in dry air at the measured room temperature, calculate the percentage error for both tuning forks [accepted value $= 331.4$ m/s $+ (0.6$ m/s/$^\circ$C$)(T_{room})$].

3. Analyze and discuss the possible sources of error in this experiment.

4. Describe how you could do a similar experiment to find the frequency of a tuning fork with an unknown frequency.

5. Was the purpose of this lab accomplished? Why or why not? (Your answer to this question should show thoughtful analysis and careful, thorough thinking.)

Invitation to Inquiry

For any sound, there is a relationship between v, f, and λ. For any sound produced in a closed air column there is also a relationship between the temperature, λ and the length of the shortest air column at which resonance occurs. Therefore, it should be possible to calibrate a closed air column, making marks on the side of the tube so you can use it as a thermometer. How can you make a sound-resonance thermometer that will show the present temperature?

Data Table 26.1   Resonance in an Air Column: Lowest Position - Frequency 1

	Trial 1	Trial 2	Trial 3	Average
Length of resonating air column (m)				
Room temperature (°C)				
Calculated Wavelength (m)				
Calculated velocity in air (m/s)				
Tuning fork frequency (Hz)...				

Data Table 26.2   Resonance in an Air Column: Next Higher Position - Frequency 1

	Trial 1	Trial 2	Trial 3	Average
Length of resonating air column (m)				
Room temperature (°C)				
Calculated Wavelength (m)				
Calculated velocity in air (m/s)				
Tuning fork frequency (Hz)...				

224

Data Table 26.3   Resonance in an Air Column: Lowest Position - Frequency 2

	Trial 1	Trial 2	Trial 3	Average
Length of resonating air column (m)				
Room temperature (°C)				
Calculated Wavelength (m)				
Calculated velocity in air (m/s)				
Tuning fork frequency (Hz)..				

Data Table 26.4   Resonance in an Air Column: Next Higher Position - Frequency 2

	Trial 1	Trial 2	Trial 3	Average
Length of resonating air column (m)				
Room temperature (°C)				
Calculated Wavelength (m)				
Calculated velocity in air (m/s)				
Tuning fork frequency (Hz)..				

Experiment 27: Reflection and Refraction

Introduction

The travel of light is often represented by a **light ray**, a line that is drawn to represent the straight-line movement of light. The line represents an imaginary thin beam of light that can be used to illustrate the laws of reflection and refraction, the topics of this laboratory investigation.

A light ray travels in a straight line from a source until it encounters some object. What happens next depends on several factors, including the nature of the material making up the object, the smoothness of the surface, and the angle at which the light ray strikes the surface. If the surface is perfectly smooth, rays of light undergo **reflection**. If the material is transparent, on the other hand, the light ray may be transmitted through the material. In these cases the light ray appears to become bent, undergoing a change in the direction of travel at the boundary between two transparent materials (such as air and water). The change of direction of a light ray at the boundary is called **refraction**.

Light rays traveling from a source, before they are reflected or refracted, are called *incident rays*. If an incident ray undergoes reflection, it is called a *reflected ray*. Likewise, an incident ray that undergoes refraction is called a *refracted ray*. In either case, a line perpendicular to the surface, at the point where the incident ray strikes, is called the *normal*. The angle between an incident ray and the normal is called the *angle of incidence*. The angle between a reflected ray and the normal is called the *angle of reflection*. The angle between a refracted ray and the normal is called the *angle of refraction*. These terms are descriptive of their meaning, but in each case you will need to remember that the angle is measured from a line perpendicular to the surface, the **normal**.

Procedure

Part A: Reflection of Light

1. Using a ruler, draw a straight line across a sheet of plain (unlined) white paper. Place the paper on a smooth piece of cardboard that has been cut from a box. Label the line with a B at one end and B′ at the other end (B is for boundary).

2. Attach a small, flat mirror to a block of wood as shown in figure 27.1. Place the mirror and block combination on the paper with the back of the mirror (the reflecting surface) on line BB′.

3. Stick a pin straight up and down into the paper about 10 cm from the mirror and slightly to the right side as shown in figure 27.1. On the left side, carefully align the edge of a ruler with the

reflected image as shown in the illustration. Then firmly hold the ruler and draw a pencil line along this edge. Move the mirror and extend this line to the mirror boundary line BB′. Label the point of reflection with the letter P.

4. Place a protractor on line BB′ and mark a point 90° from the line. From this point, use the ruler to draw a dashed normal (NP). Complete your ray diagram by using the ruler to draw a line from the point of reflection (P) to the source of the light ray at the pin (I). Place arrows on line IP and line PR to show which way the light ray moved.

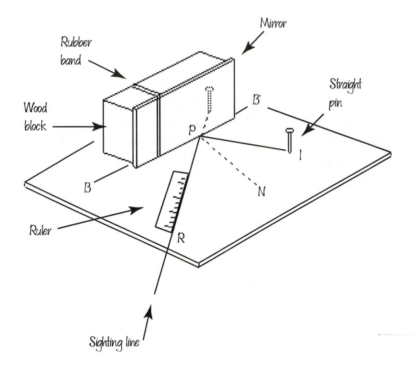

Figure 27.1

5. Use the protractor to measure the angle of incidence and the angle of reflection. Record these angles in Data Table 27.1 under Trial 1.

6. Place the mirror with its back edge again on line BB′ and conduct a second and third trial at different sighting angles. Record these measurements in Data Table 27.1 on page 232.

Part B: Refraction of Light

1. Place a clean sheet of white (unlined) paper on the cardboard. Place a glass plate approximately 5 cm square flat on the center of the paper. Use a pencil to outline the glass plate, then move the plate aside.

228

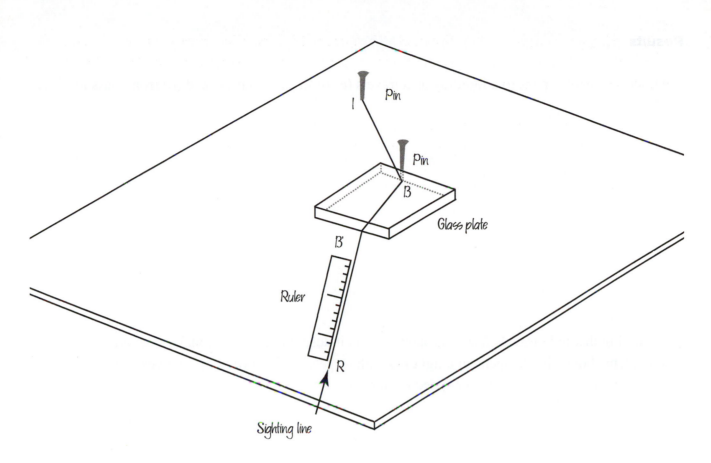

Figure 27.2

2. Use a ruler to draw a straight line from the upper left edge of the plate outline, making an angle of about 60° to the edge. Label this line IB as shown in figure 27.2. Place one upright pin at point B immediately outside the plate outline and a second upright pin at point I. Return the glass plate to the outline.

3. Bring the cardboard, paper, and glass plate near the edge of the table so you can sight through the glass plate toward the two pins. Position a ruler so that one edge aligns with the two pins as shown in figure 27.2. Draw a line along the ruler and label the line B′R. Move the glass plate aside for a second time.

4. Draw a line from B to B′, showing the path of the light ray through the glass. Overall, the path of the light ray is from IB to BB′ to B′R, showing that the light ray was bent twice.

5. Draw normals to the surface of the glass at B and B′. Show the angle of incidence and the angle of refraction with curved arrows at both boundaries.

Results

1. Describe what happens to a light ray as it travels (a) from air into glass and (b) from glass into air.

2. Assuming that light travels faster through air than it does through glass, make a generalized statement about what happens to a light ray with respect to the normal as it moves from a faster speed in one material to a slower speed in another.

3. What rules or generalizations do your findings suggest about reflection? How much more data would be required to make this a valid generalization?

4. Does the rule your findings suggest about reflection apply in all types of reflection? Explain.

5. What rules or generalizations do your findings suggest about refraction? How much more data would be required to make this a valid generalization?

6. Was the purpose of this lab accomplished? Why or why not? (Your answer to this question should be reasonable and make sense, showing thoughtful analysis and careful, thorough thinking.)

Invitation to Inquiry

Trace the travel of light rays through a convex lens, a concave lens, and a triangular prism. What are the factors that determine how much the light ray is bent in each lens?

Data Table 27.1	Reflection of Light	
Trial	Angle of Incidence	Angle of Reflection
1	_____	_____
2	_____	_____
3	_____	_____

Experiment 28: Physical And Chemical Change

Introduction

Matter undergoes many changes and most of the common, everyday changes are either physical or chemical. A **physical change** is one that does not alter the identifying properties of a substance. It can be a change in form, state, or energy level, but no permanent change occurs in the properties of the substance. A piece of paper torn into two parts, for example, still has the properties of the original paper and no new substance has been formed. Thus tearing a piece of paper into two parts is a physical change. Evaporation, condensation, melting, freezing, and dissolving often produce physical changes.

A **chemical change** is one that produces new substances with new properties. A piece of paper that burns produces gases and ash that have different properties than the original paper, so this is an example of a chemical change. Heat, light, and electricity often produce chemical changes. Chemical changes occur as (1) atoms combine to form new compounds, (2) compounds break down into simpler substances, and (3) compounds react with other compounds or elements to form new substances. In this experiment you will determine if certain changes in matter are physical or chemical changes.

Materials

Sodium chloride, graduated cylinder, evaporating dish, nichrome wire, tongs, magnesium ribbon, small test tube, dilute hydrochloric acid, silver nitrate solution, funnel, funnel support, ring stand, filter paper, copper(II) chloride solution, beaker, aluminum foil.

Procedure

1. Dissolve about 2 g of sodium chloride (ordinary table salt) in 50 mL of water in a clean graduated cylinder. Observe (a) if the water level changes when the salt is added, and (b) the taste of the solution. **CAUTION:** Taste chemicals only when directed to do so. Record your observations in Data Table 28.1.

2. Continue dissolving 2 g samples of sodium chloride until 10 g are dissolved. Record your observations about the water level and the taste after each of the five additions.

3. Place about 5 mL of the solution in an evaporating dish and heat slowly until a dry solid remains. When the solid has cooled, carefully taste the solid and record your observation in the data table.

4. Carefully examine a 5-cm length of nichrome wire, noting the color, luster, flexibility, and other properties you can observe. Hold the wire with tongs and heat it strongly in the flame of a burner until it glows brightly. After the wire cools, again examine the wire, comparing the properties before and after heating. Record your observations.

5. Examine a 5-cm length of magnesium ribbon, noting the color, luster, flexibility, and other properties you can observe. Hold the ribbon with tongs and ignite the end in the flame of a burner. **CAUTION**: Do not look directly at the magnesium while it is burning. Compare the properties of the ash with the properties of the original magnesium ribbon, recording your observations in the data table.

6. Pour about 5 mL of silver nitrate solution in a small test tube. Add several drops of dilute hydrochloric acid, gently swirling the mixture with the addition of each drop. Record your observations in the data table. Filter the solid that forms (the precipitate), using small amounts of water as necessary to flush all the precipitate from the test tube. (Figure 28.1 shows how to fold a filter paper and figure 28.2 shows how to set up apparatus to filter a liquid.) Discard the filtered liquid (the filtrate) and place the filtered precipitate in direct sunlight. Record your

Step 1: Fold along diameter.

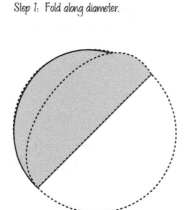

Step 2: Fold over a second time.

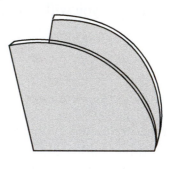

Step 3: Open one fold to make cone.

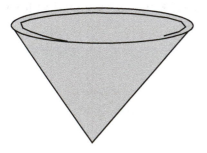

Step 4: Place cone in funnel.

Figure 28.1

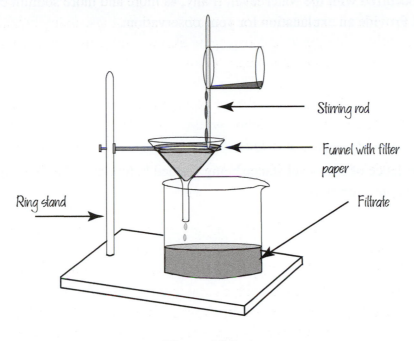

Figure 28.2

observations after the precipitate has been in the sunlight for two or three minutes.

7. Pour about 125 mL of copper(II) chloride solution in a beaker. Observe if the beaker feels cool or warm to touch. Obtain a piece of aluminum foil approximately 2 cm by 15 cm and form it into a loose coil. Drop the aluminum foil coil into the copper(II) chloride solution. Observe the results, if any, and cautiously touch the beaker after about five minutes. Record your observations in the data table.

Results

1. What kind of change occurs when sodium chloride dissolves? Give an explanation for your answer.

2. What changes occurred with the water level, if any, as more and more sodium chloride was added to the solution? Provide an explanation for your observation.

3. Compare the evidence of a new substance being formed before and after heating (a) a nichrome wire, and (b) a magnesium ribbon.

4. What evidence indicates if new substances with new properties were or were not formed when hydrochloric acid was added to a silver nitrate solution?

5. Did the precipitate exposed to sunlight undergo either a chemical or physical change? Give an explanation for your answer.

6. What changes did you observe when the aluminum was added to the copper(II) chloride solution? (Hint: There were more than five changes.)

7. Which of the changes in this experiment were physical changes? Give reasons for your conclusion.

8. Which of the changes in this experiment were chemical changes? Give reasons for your conclusion.

9. Was the purpose of this lab accomplished? Why or why not? (Your answer to this question should be reasonable and make sense, showing thoughtful analysis and careful, thorough thinking.)

Invitation to Inquiry

In many communities the recycling of aluminum, paper, and plastics is started by first segregating items made of these materials from the rest of the trash. A major problem in recycling plastics is the many different types of plastics that exist, all with different chemical and physical properties. Some of these materials are more desirable for recycling than others, so they must be sorted. One way of sorting plastics is to read the code that might be stamped on the bottom. Here are some letter and number codes from some common plastic items. The number usually appears inside the recycling arrow logo: 2 (HDPE) milk jugs, bleach and detergent bottles; 1 (PETE) soft-drink bottles; 5 (PP) ketchup bottles, yogurt cups; 6 (PS) transparent plastic drinking cups, CD boxes, and; 4 (LDPE) plastic squeeze bottles. Can you find a way to separate a mixture of pieces of plastic from each of these 5 groups by taking advantage of the differences in chemical or physical properties? Consider cutting pieces of plastic from each of the group and finding important properties that could be used in a separation scheme.

Data Table 28.1	Physical and Chemical Changes
Action	Observations
Dissolving sodium chloride	
Evaporating sodium chloride solution	
Heating nichrome wire	
Heating magnesium ribbon	
Silver nitrate and hydrochloric acid mixture	
Precipitate in sunlight	
Aluminum foil in copper(II) chloride solution	

Experiment 29: Hydrogen

Introduction

Hydrogen is the most abundant element in the whole universe, making up the bulk of the stars and the matter found in space between the stars. However, hydrogen is relatively scarce on earth and only trace amounts are found in the earth's atmosphere. Most of the hydrogen atoms found the earth are (1) combined with oxygen in the form of water, and (2) combined with organic compounds in living things and in fossil fuels. Very little hydrogen is found in other compounds. It is theorized that the earth originally had an atmosphere of abundant hydrogen, but hydrogen is the lightest of all the gases and was easily swept away early in the earth's history. The present-day atmosphere evolved after the hydrogen was removed.

There are no readily available sources of hydrogen as an element in the atmosphere or from the earth's surface. It can be obtained from the decomposition of water and other compounds. In the past, hydrogen obtained from chemical reactions was used to fill balloons and blimps because it is the least dense of all the gases. Because of its flammability, hydrogen gas has been replaced by the slightly more dense but inflammable helium gas. Today, hydrogen is used primarily as a fuel, in the commercial production of ammonia, and to hydrogenate unsaturated fats. As a fuel, hydrogen combines with oxygen to produce water, releasing a large amount of heat. A temperature of 2500° C, for example, is reached in the flame of an oxy-hydrogen torch. The reaction is

$$2 H_2 + O_2 \rightarrow 2 H_2O.$$

Laboratory preparation of hydrogen is usually accomplished by the displacement of hydrogen from an acid by an active metal. Zinc, for example, may be used to replace hydrogen from sulfuric acid.

CAUTION: Hydrogen mixed with air is explosive. Keep flames away from all hydrogen generators, which must have tightly fitting parts.

Materials

About 20 g mossy zinc, dilute sulfuric acid (1 vol. concentrated H_2SO_4 to 4 vol. water), long wood splints, apparatus shown in figure 29.1, three small collecting bottles (250 mL or less), three glass plates, copper(II) sulfate solution, small test tube.

Procedure

Part A: Generation of Hydrogen

1. Set up the hydrogen generator as shown in figure 29.1. Place about 20 g mossy zinc in the generator bottle. Make sure the funnel tube (or thistle tube) extends very close to the bottom of the generator bottle. The bottom of the tube must extend well into the acid solution that will be added. Fill three small collecting bottles with water and invert them in a half-filled pneumatic trough. Your instructor should approve your apparatus setup *before* you proceed to the next step.

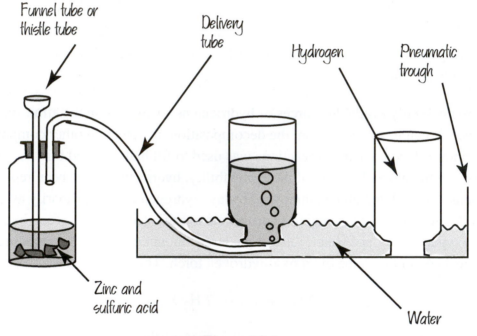

Figure 29.1

2. After your apparatus has been approved, slowly add dilute sulfuric acid through the funnel tube, making sure the bottom of the tube is below the liquid level. If the reaction proceeds slowly, add several drops of copper sulfate through the funnel tube. **CAUTION**: Keep all flames well away from the hydrogen generator.

3. Collect a *small* test tube of hydrogen gas by water displacement. Carry it *mouth downward* away from the generator to a flame. If it burns with a gentle puff, you are collecting pure hydrogen. If it burns with a pop, it is still impure. Continue testing the hydrogen with small test tube amounts until you are collecting pure hydrogen. Then fill the three bottles by water displacement, leaving each bottle upside down in the pneumatic trough. In this experiment, leave the delivery tube under water in the trough when you have collected the hydrogen gas. Add 100 mL of water through the funnel tube to stop the reaction.

Part B: Properties of Hydrogen

1. Place a glass plate over the mouth of one bottle of hydrogen gas while it is still under water. With the plate in place, carry the bottle mouth downward to a location well away from any hydrogen generators. Stand the bottle mouth upward and remove the plate. Wait several minutes, then bring a flaming splint to the mouth of the bottle. Record your observations in Data Table 29.1.

2. Place a second bottle of hydrogen gas on a glass plate while it is still under water, and again carry the bottle mouth downward to a location well away from any hydrogen generators. Keeping the bottle mouth downward, slowly remove the glass plate and bring a long flaming wood splint to the mouth of the bottle. Move the flame with the bottle and slowly turn the bottle mouth upward. Record all observations, including any changes you note on the inside walls of the bottle.

3. As before, carry the third bottle of hydrogen gas mouth downward, covered with a glass plate, well away from the hydrogen generators. Keeping the bottle mouth downward, remove the glass plate and insert a long flaming wood splint well inside the bottle. Withdraw the splint and reinsert it several times. Record your observations in the data table.

Results

1. Why is a bottle of hydrogen gas carried mouth downward?

2. What physical property of hydrogen gas explains the observations of what happens when a flame is brought near an upright bottle of hydrogen gas? Explain.

241

3. Considering the mixing of hydrogen gas with air, explain the meaning of what happened when a bottle of hydrogen gas was turned upright with a lighted splint near the mouth.

4. Where does combustion take place when a flaming splint is inserted and withdrawn from the mouth of a mouth–down bottle of hydrogen gas? Offer an explanation for this observation.

5. What is formed from the combustion of hydrogen?

6. What are the physical properties of hydrogen gas?

7. Is hydrogen gas soluble in water? Provide experimental evidence for your answer.

8. Write word or balanced chemical equations that describe the following reactions:

      (a) zinc plus sulfuric acid

      (b) hydrogen combining with oxygen

9. What are some chemical properties of hydrogen gas? Support your answer with experimental evidence.

Invitation to Inquiry

     Are the laboratory results obtained always what you expect? If not, what is the meaning of unexpected results? Measure exactly 50 mL water into a 100 mL graduate cylinder. Carefully measure 50 mL of ethyl alcohol (pure as possible) into another graduated cylinder. Place both graduated cylinders on a balance and record the mass. Slowly pour the alcohol into the water. Note the combined volume and the total mass. Were the findings expected? What could possibly explain this?

Data Table 29.1	Properties of Hydrogen
Action	**Observations**
Flaming splint brought to mouth of upright bottle	
Flaming splint brought to a mouth-down bottle rotated upright	
Flaming splint in and out of a mouth-down bottle	

Experiment 30: Oxygen

Introduction

Oxygen is the most abundant element at the surface of the earth. The atmosphere contains almost 21 percent oxygen by volume. On the surface, water is the most abundant compound and water is 89 percent oxygen by weight. The outer solid part of the earth's crust is made up of silica and other oxygen-containing minerals, with oxygen contributing about 47 percent of the total. Altogether, oxygen makes up about 50 percent of the earth's atmosphere, waters, and solid crust.

Many compounds that contain oxygen decompose when heated, releasing oxygen. For example, the oxides of less active metals such as mercury, silver, gold, and platinum all release oxygen when heated. Oxygen was first obtained in 1774 when Joseph Priestley focused sunlight on mercury(II) oxide with a magnifying glass:

$$2\,HgO \xrightarrow{\Delta} 2\,Hg + O_2 \uparrow$$

Peroxides decompose when heated or when placed in the presence of certain catalysts. Hydrogen peroxide is a common example:

$$2\,H_2O_2 \xrightarrow{\Delta} 2\,H_2O + O_2 \uparrow$$

Most oxygen used for industry and commercial purposes is prepared by the fractional distillation of liquid air. Liquid nitrogen and argon boil away from liquid air between $-186°$ and $-196°$ C, leaving nearly pure oxygen. Liquid oxygen has a boiling point of $-183°$ C.

Oxygen is commonly prepared in the laboratory by the decomposition of potassium chlorate in the presence of a catalyst, such as manganese dioxide:

$$2\,KClO_3 \xrightarrow[MnO_2]{\Delta} 2\,KCl + 3\,O_2 \uparrow$$

CAUTION: Impurities in the potassium chlorate can produce an explosion. Make sure that all utensils are clean and that no foreign materials are introduced in the prepared potassium chlorate and manganese dioxide mixture.

Materials

Mixture of two parts of potassium chlorate ($KClO_3$) and one part of manganese dioxide (MnO_2), large Pyrex test tube and other apparatus shown in figure 30.1, four 500-mL collecting bottles, four glass plates, wood splints, limewater, charcoal stick, forceps, steel wool.

Procedure

Part A: Generation of Oxygen

1. Set up the oxygen generator as shown in figure 30.1. (If your room does not have sinks, place the pneumatic trough in a large pan or tray to catch water overflow.) Place about 5 g of the potassium chlorate-manganese dioxide mixture in the large Pyrex test tube.

2. Fill four collection bottles with water and cover each with a glass plate. Fill the pneumatic trough about half full of water. Place each bottle upside down in the trough, being careful not to introduce any air bubbles into the bottles. Remove the glass plates and set them aside for now. Your instructor should approve your apparatus setup *before* you proceed to the next step.

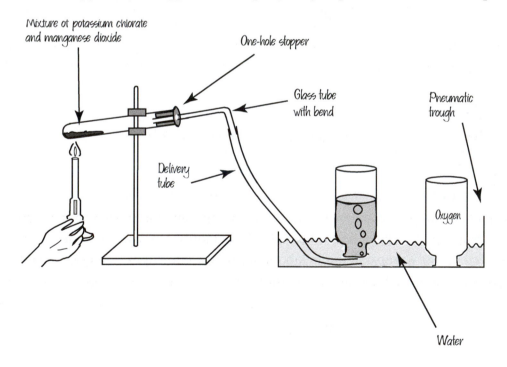

Figure 30.1

3. Light the Bunsen burner and hold it by the base. Gently heat the tube, starting at the top of the mixture and constantly moving the flame back and forth. You are trying to produce a slow but

246

steady stream of bubbles without overheating the tube. If any of the mixture starts to move upward, gently tap the tube. Allow the bubbles to rise in the water for about 10 or 15 seconds after a steady stream of bubbles begin. Then place the mouth of one of the upside down bottles over the end of the delivery tube. If the bubbles start streaming too rapidly, or if a white vapor appears, you are heating the mixture too strongly. If this happens remove the flame from the tube for a moment, then continue slowly moving the flame back and forth across the part of the tube containing the mixture.

4. When all of the water has been displaced from one bottle, carefully slide it aside and fill another bottle. When all four bottles are filled, remove the delivery tube from the water *before* you turn off the flame.

5. Hold a glass plate securely against the mouth of an upside down bottle while it is under water, remove it from the trough, then set it upright on the table. Keep the bottles covered with glass plates until you are ready to use them in the next part of this experiment.

Part B: Properties of Oxygen

1. Observe the color of the oxygen gas you have collected in the bottles. Uncover one of the bottles and gently waft some of the gas toward your nose, cautiously sniffing to observe any odor. Record your observations in Data Table 30.1.

2. Light a wood splint with a match and allow it to burn for a few seconds. Blow out the flame so a glowing spark remains. Quickly remove a glass plate from a second bottle of oxygen gas, hold the bottle at a slant and insert the glowing splint into the gas. Dip the splint in water, then record your observations in the data table.

3. Hold a stick of charcoal, which is mostly carbon, with tongs and place one end in a burner flame until it glows. Note how the charcoal burns in air, then lower it into a third bottle of oxygen gas while keeping the mouth mostly covered with the glass plate. Remove the charcoal and douse it with water, leaving the glass plate over the bottle. Add a small amount of clear limewater, then hold the glass plate over the bottle and shake it thoroughly. Record how the charcoal burns in air and in oxygen, and what happens to the limewater.

4. Add several centimeters of water to the fourth bottle of oxygen gas and quickly replace the glass plate. Using tongs, heat a small wad of steel wool in a burner flame until it sparkles. Quickly remove the glass plate and hold the steel wool inside the bottle. Do not allow the steel wool to touch the sides of the bottle. Record your observations in the data table.

Results

1. Describe what happens when potassium chlorate is heated.

2. What is the purpose of the manganese dioxide in the oxygen generation mixture?

3. In generating oxygen gas: (a) Why were bubbles allowed to rise for 10 to 15 seconds before the gas was collected? (b) Why was the delivery tube removed before the flame was turned off?

4. Account for your observations of what happened when a glowing splint was placed in oxygen gas as compared to air.

5. What product was formed when (a) burning charcoal (carbon), and (b) burning steel wool (iron) were placed in oxygen? Provide evidence for your answers in each case.

6. What are the physical properties of oxygen gas?

7. Write word or balanced chemical equations that describe

(a) potassium chlorate heated;

(b) carbon combining with oxygen; and

(c) iron combining with oxygen if the product is Fe_3O_4.

8. Does oxygen burn or does it support combustion? Support your answer with experimental evidence.

Note: *Flammable* and *inflammable* both mean easily ignited and capable of burning quickly. Flammable has been adopted for the labeling of combustible materials because some people thought the in- of inflammable meant "not."

Invitation to Inquiry

Oxygen gas was collected in this laboratory investigation by bubbling the gas through water. Does much of the gas dissolve in water during the process? If so, is the temperature of the water important in determining how much gas dissolves in the water? You can test these relationships by setting up an experiment with Alka-Seltzer® tablets. Set up a flask on a sensitive balance, with two tables wrapped in a tissue, then lodged inside the neck of the flask. Record the weight, then gently push the tissue so it falls into the water. The difference in mass will be a result of carbon dioxide leaving the flask. Compare how much carbon dioxide is dissolved in water of different temperatures. Should you expect the same dissolving rate for oxygen?

Data Table 30.1   Properties of Oxygen

Action	Observations
Observe color and odor	
Insert glowing splint	
Burning charcoal observed: (a) In air (b) In oxygen gas	
Limewater added to bottle after burning charcoal	
Burning steel wool observed: (a) In air (b) In oxygen gas	

Experiment 31: Conductivity of Solutions

Introduction

Ionic compounds that dissolve in water do so by ions being pulled from the crystal lattice to form a solution of free ions. When free ions (charged particles) are present in a solution, the solution is a good conductor of electricity. Many **covalent compounds** that dissolve in water form molecular solutions of noncharged particles. Such a solution will not conduct an electric current. Some covalent compounds are pulled apart into ions; that is, the compound is *ionized* in a water solution. Strong ionization results in a solution that is a good conductor of electricity. Partial ionization results in a solution that is a poor conductor of electricity. The amount of current that flows through such a solution is roughly related to the amount of ionization. In this experiment you will use a conductivity apparatus (figure 31.1) to compare the conductivity of different solutions.

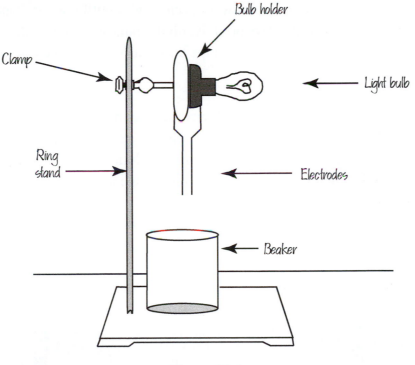

Figure 31.1

CAUTION: The solutions in this experiment are dilute and relatively safe provided that you exercise care in their use. Acids and bases, however, can irritate the skin and damage clothing and other materials. If you spill an acid or base on your skin or clothing, wash it off immediately with plenty of water and inform your instructor.

Materials

      Conductivity apparatus (figure 31.1), short patch cord with alligator clips, distilled water, tap water, dry sodium chloride (table salt), sugar solution (1%), ethyl alcohol, hydrochloric acid (1.0 M: 85 mL concentrated/L solution), sodium hydroxide (1.0 M: 40 g solid/L solution), sodium chloride solution (1.0 M: 58 g solid/L solution), vinegar, glycerin.

Procedure

1. Set up the apparatus as shown in figure 31.1. With the lamp *unplugged*, attach a patch cord with alligator clips to the two electrodes. Plug in the lamp to make sure the apparatus works, then unplug the lamp. Remove the patch cord and set it aside.

2. Test each substance listed in Data Table 31.1, using the conductivity apparatus with the following procedure:

    For solids: With the lamp *unplugged*, lower the electrodes until both are touching the solid. Plug in the lamp and record if it glows brightly, dimly, faintly, or not at all. *Unplug* the lamp and remove the electrodes from the solid. Wipe the electrodes with a clean cloth before proceeding.

    For solutions: With the lamp *unplugged*, lower the electrodes to the same depth in each solution. Plug in the lamp and record if the bulb glows brightly, dimly, faintly, or not at all. *Unplug* the lamp and remove the electrodes from the solution. Wash the electrodes by immersing them in distilled water before proceeding to the next solution test.

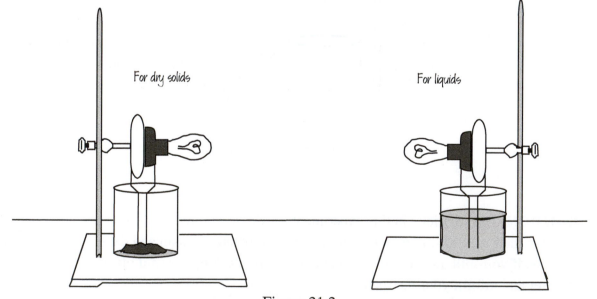

Figure 21.2

3. Your instructor might specify other solids or solutions to be tested, which should be added to the list in Data Table 31.1.

Results

1. Compare your observations of the conductivity of dry sodium chloride and a solution of sodium chloride. What conclusions can you make about these observations?

2. Compare your observations of the conductivity of distilled water and tap water. Offer an explanation for this comparison.

3. Compare the conductivity of hydrochloric acid and vinegar, which is also an acid solution. If the amount of conductivity is an indication of the degree of ionization of an acid solution, what would you predict about the degree of ionization of these two solutions?

4. What solutions were good conductors of electricity? What solutions were nonconductors of electricity? Describe any patterns that you can find in this summary.

5. In general, what kind of compounds conduct electricity when they are in solution? Explain.

6. In general, what kind of compounds do *not* conduct electricity when they are in solution? Is this always true? Explain.

7. Was the purpose of this lab accomplished? Why or why not? (Your answer to this question should be reasonable and make sense, showing thoughtful analysis and careful, thorough thinking.)

Invitation to Inquiry

Use some small gauge wire, a flashlight battery, and a flashlight bulb and holder to set up your own portable conduction tester. Test a variety of materials to find which are conductors and which are nonconductors. Generalize what categories of materials are conductors and what categories seem to be nonconductors.

Data Table 31.1   Conductivity Tests of Solids and Solutions

Substance	Observations
Dry solid sodium chloride	
Distilled water	
Tap water	
Sodium chloride solution	
Hydrochloric acid	
Vinegar	
Ethyl alcohol	
Sodium hydroxide solution	
Glycerin	
Sugar solution	

Experiment 32: Percentage Composition

Introduction

A chemical formula is a shorthand way of describing the elements or ions that make up a compound. For covalent compounds, the chemical formula identifies the actual numbers of atoms in a molecule and is known as a **molecular formula**. For example, ordinary table sugar (sucrose) is a covalent compound that exists as molecules made up of carbon (C), hydrogen (H), and oxygen (O) atoms. The molecular formula for table sugar is $C_{12}H_{22}O_{11}$, which tells you that there are 12 carbon atoms combined with 22 hydrogen atoms and 11 oxygen atoms. When strongly heated, the sugar molecule decomposes to water and carbon. The water evaporates, leaving a residue of black carbon.

The **molecular weight** of a substance is the sum of the atomic weights of all the atoms in the molecule. It is the average mass of a molecule expressed in atomic mass units (u). A molecular weight can be determined experimentally in the laboratory. If the molecular formula is known, however, it can be used to determine the **formula weight**. The formula weight is the sum of the atomic weights of all the atoms in the formula of a substance. The molecular weight and a calculated formula weight are identical for a molecular substance. The formula weight of sugar can thus be determined from a periodic table. According to the table the approximate (rounded) atomic weight of carbon is 12 u, oxygen is 16 u, and hydrogen is 1.0 u. Table 32.1 shows how to add the atomic weights for all the atoms in a molecule of $C_{12}H_{22}O_{11}$.

Atoms	Atomic Weight	Totals
12 of C	12×12 u $=$	144 u
22 of H	22×1 u $=$	22 u
11 of O	11×16 u $=$	176 u
	Formula weight $=$	342 u

Table 32.1

The formula weight of sugar (342 u) can be used to find the *theoretical* percentage of water in a sample of $C_{12}H_{22}O_{11}$. The theoretical percentage is found by dividing the atomic weight of the hydrogen and oxygen atoms in the molecule by the formula weight.

In this experiment, you will determine the percentage of water in a sample of table sugar. The *experimental* percentage of water is found by measuring the mass of a sample, heating it strongly to drive off all the water, then measuring the mass of the carbon residue. The difference in mass is the mass of water that was in the original sample. By dividing this mass by the mass of the original sample, you can calculate the percentage of water in the original compound.

CAUTION: The odor of burnt sugar is emitted during this experiment. The room should be well-ventilated or the sugar should be heated in a fume hood.

Materials

Test tube (to be discarded after the experiment), test tube holder, burner, balance, desiccator (or large jar with lid partly filled with anhydrous calcium chloride), sucrose (table sugar).

Procedure

1. Heat a clean, dry test tube gently for three or four minutes to evaporate any water and remove any impurities that might be present. Cool the test tube in a desiccator. While the test tube is cooling, measure about 2 g of table sugar on a square of paper.

2. Accurately measure the mass of the still-warm test tube to the nearest centigram and record the mass in Data Table 32.1. Hold the paper square with the sugar on it in your hand and carefully make a slight crease in the paper. Slide the sugar down this crease into the test tube. Measure the mass of the test tube and sugar accurately to the nearest centigram. Record the mass in the data table.

3. Heat the sugar *very gently* for several minutes, observing any changes in the sugar and any changes around the cooler upper part of the test tube. Gradually increase the heat, heating the contents until no more changes are observed. Then heat the test tube very strongly until the contents appear dry.

4. Cool the test tube in a desiccator. Review the calculations to be done while the test tube is cooling. Measure the mass of the warm test tube and contents accurately to the nearest centigram. Record this measurement as Trial 1.

5. To ensure that all the water has been driven off, repeat procedure steps 3 and 4 for a second trial. Continue with more trials as necessary until a consistent mass is obtained; that is, until a new trial measurement is the same as the previous trial. Record the consistent mass of the test tube and residue in the data table.

6. Calculate the experimental percentage of water lost according to the experimental data.

7. Calculate the theoretical percentage of water according to the formula weight, which is the accepted value, then find the percentage error.

Results

1. What did you observe as a sample of sugar was heated? What would account for these observations?

2. Why does the use of a desiccator improve the results of this experiment?

3. Why was the residue reheated for several trials?

4. Analyze the possible sources of error in this experiment.

5. Was the purpose of this lab accomplished? Why or why not? (Your answer to this question should be reasonable and make sense, showing thoughtful analysis and careful, thorough thinking.)

Invitation to Inquiry

Epsom salts is the common name for magnesium sulfate heptahydrate, which contains water. The water is released when a sample of Epsom salts is heated. Most drug store chains have their own generic brand of Epsom salts. Do the different brands have the same proportion of water? Using the procedures of this lab investigation, you should be able to experimentally verify if all brands of Epsom Salts have the same proportion of water. Test several different brands of Epsom salts, then report your findings to your class.

Data Table 32.1   Percentage of Water in Sucrose

1.	Mass of dry test tube -----------------------------------	_____ g
2.	Mass of test tube and sugar before heating -----------	_____ g
3.	Mass of sugar before heating (row 2 – row 1) -----	_____ g
4.	Constant mass of test tube and residue --------------	_____ g
5.	Mass of residue only (row 4 – row 1) ---------------	_____ g
6.	Mass of water driven off (row 3 – row 5) ----------	_____ g
7.	Percentage of water lost (row 6 ÷ row 3 × 100) ---	_____ %
8.	Theoretical percentage of water ------------------------	_____ %
9.	Percentage of error -------------------------------------	_____ %

Experiment 33: Metal Replacement Reactions

Introduction

Based on what happens to the reactants and products, there are basically four types of chemical reactions: (1) combination, (2) decomposition, (3) replacement, and (4) ion exchange reactions. This experiment is concerned with **metal replacement reactions**. A metal replacement reaction occurs when a more active metal is added to a solution of the salt of a less active metal. In generalized form the reaction is XY + A → AY + X, where A represents the more active metal and XY represents the salt of a less active metal.

A metal replacement reaction occurs because some metals have a stronger electron holding ability than other metals. Metals that have the least ability to hold on to their electrons are the most chemically active. An activity series ranks the most chemically active metals at the top and the least chemically active at the bottom. This means that the activity series is upside down with respect to electron holding ability. Metals at the top of the activity series have the least ability to hold on to their electrons and those at the bottom have the greatest ability to hold on to their electrons.

A metal replacement reaction occurs when a more active metal (with less electron holding ability) is added to a solution containing the ions of a less active metal (with greater electron holding ability). The more active metal loses electrons to the less active metal, so they trade places; that is, the active metal loses electrons to form metallic ions in solution, and the less active metal gains electrons and comes out of solution as a solid metal. In this experiment you will study the chemical reaction of three metals and rank them according to their electron holding ability.

CAUTION: Soluble lead compounds are poisonous. Silver nitrate solutions will stain the skin. Avoid contact with both of these metal salt solutions. Wash thoroughly with soap and water if contact is suspected. Wash your hands after completing the experiment.

Materials

Metal strips (about 2 cm × 5 cm) of copper, zinc, and lead; sandpaper; test tubes; test tube rack; graduated cylinder; 10-cm length of thin copper wire; silver nitrate solution (0.1 M: 17.0 g/L of solution); copper nitrate solution (0.1 M: 24.2 g/L of solution); lead nitrate solution (0.1 M: 33.1 g/L of solution).

Procedure

1. Obtain three clean, dry test tubes and a test tube rack. Add 10 mL of silver nitrate solution to each test tube.

2. Obtain three metal strips each of copper, zinc, and lead. Sandpaper all nine strips until they are clean and bright. Make a right-angle bend near the end of each metal strip. A 10-cm length of thin copper wire with one end formed into a hook can be used to fish the bend of a metal strip from a test tube. Use the copper wire to remove the metal strips from their solution for inspection as needed.

3. Place a strip of each metal into a test tube with the silver nitrate solution. Observe the metals for 10 minutes or so, looking for evidence of a replacement reaction. Carefully touch each test tube at the outside bottom to observe any temperature changes. If a black deposit appears on a metal strip, observe for an additional 5 minutes or longer. In Data Table 33.1, record your observations of any changes in the solutions and any changes on the metal strips.

4. Repeat procedure steps 1 through 3 with fresh pieces of metal and 10 mL of copper nitrate solution. Observe the metals for 10 minutes or more, again looking for evidence of a replacement reaction.

5. Repeat procedure step 3 again, this time with fresh pieces of metal and 10 mL of lead nitrate solution. Again observe the metals for 10 minutes or longer and record your observations in the data table.

Results

1. According to Data Table 33.1, which metals had the (a) greatest number of reactions; (b) middle number of reactions; (c) least number of reactions?

2. Are your findings in question 1 consistent with the activity series of metals? Explain.

3. Were any changes observed in the color of the solutions? Offer an explanation for this observation.

4. Complete the following equations, writing "no reaction" if none occurred:

$$Cu + AgNO_3 \rightarrow$$

$$Zn + AgNO_3 \rightarrow$$

$$Pb + AgNO_3 \rightarrow$$

$$Cu + Cu(NO_3)_2 \rightarrow$$

$$Zn + Cu(NO_3)_2 \rightarrow$$

$$Pb + Cu(NO_3)_2 \rightarrow$$

$$Cu + Pb(NO_3)_2 \rightarrow$$

$$Zn + Pb(NO_3)_2 \rightarrow$$

$$Pb + Pb(NO_3)_2 \rightarrow$$

5. Write a general rule describing when metal replacement reactions occur and when they do not occur.

6. Was the purpose of this lab accomplished? Why or why not? (Your answer to this question should be reasonable and make sense, showing thoughtful analysis and careful, thorough thinking.)

Invitation to Inquiry

Make a plan for using the fewest number of steps possible in an experimental study to place six metals (Mg, Zn, Fe, Sn, Cu, and Al) in an activity series from highest to lowest. Describe the procedure you would use for determining the position of a Ni salt in this series.

Data Table 33.1   Metal Replacement Reactions

	Zinc metal	Copper metal	Lead metal
Silver nitrate solution			
Copper nitrate solution			
Lead nitrate solution			

Experiment 34: Producing Salts by Neutralization

Introduction

An acid can be considered as a solution of hydronium ions (H_3O^+) and some negative ions. All acids furnish hydronium ions in a water solution and the type of negative ions present depend on which acid is dissolved in the water. A hydrochloric acid solution, for example, is a water solution of positive hydronium ions and negative chloride ions. A base, on the other hand, can be considered as a solution of positive metal ions (or ammonium) and negative hydroxide ions (OH^-). All bases furnish hydroxide ions in a water solution and the type of positive ions formed depends on which base is dissolved in the water. A solution of calcium hydroxide, for example, is a water solution of positive calcium ions and negative hydroxide ions.

When an acid solution and a base solution are mixed, the hydronium ions of the acid unite with the hydroxide ions of the base in an ionic exchange reaction. A molecule of water is formed, removing the ions characteristic of acids and bases. Since these ions are removed, the acid and base properties no longer exist and the acid and base are said to be **neutralized**. When an acid and base neutralize each other, the products formed are water and a salt.

A **salt** is defined as any ionic compound except those with hydroxide or oxide ions. As an example of a salt produced by a neutralization reaction, consider the reaction of a solution of hydrochloric acid with a solution of calcium hydroxide. The reaction is

$$2\ HCl + Ca(OH)_2 \rightarrow CaCl_2 + 2\ H_2O.$$

Molecular water is formed in this reaction, leaving calcium ions (Ca^{+2}) and chloride ions (Cl^-) in solution. As the water is evaporated, these ions begin forming ionic crystal structures as the concentration of the solution increases. When all of the water evaporates, the white crystalline salt of calcium chloride ($CaCl_2$) remains.

Different kinds of salts can be prepared by using different kinds of acids with different kinds of bases. In this experiment you will prepare two different salts by neutralization reactions.

Materials

Evaporating dish, small beakers, dropper, phenolphthalein solution, graduated cylinder, glass stirring rod, universal indicator paper (with color-coded pH scale), watch glass, ring stand and ring, wire screen, burner, hydrochloric acid (1.0 M: 81 mL concentrated/L of solution), sulfuric acid (1.0 M: 56 mL concentrated/L of solution), sodium hydroxide solution (1.0 M: 40 g/L of solution), potassium hydroxide solution (1.0 M: 56 g/L of solution), distilled water.

CAUTION Acids and bases can damage skin and other materials. If spilled on your person or clothing, flush with running water for several minutes and inform the instructor. Dilute any tabletop spills with water, then use a sponge to wipe up the spill. Be sure to rinse the sponge with plenty of water.

Procedure

Part A: Preparation of Sodium Chloride

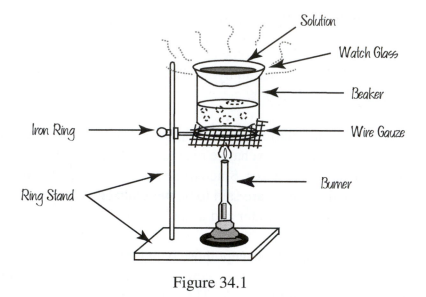

Figure 34.1

1. Place two strips of universal indicator paper in a clean watch glass. Pour 5 mL of the sodium hydroxide solution into a labeled beaker. Pour 5 mL of the dilute hydrochloric acid solution into a second labeled beaker. Use a clean, dry glass stirring rod to transfer a drop of each to the indicator paper. Match the color change caused by each with the color standard on the containers, then record the pH of each in Data Table 34.1.

2. Pour the 5 mL of sodium hydroxide solution into a clean evaporating dish. Add a drop of phenolphthalein indicator solution to the evaporating dish. Note the color of the indicator in the base.

3. Slowly add about half of the hydrochloric acid solution to the dish, carefully stirring with a dry, clean glass stirring rod. Use a medicine dropper to add more hydrochloric acid, a drop at a time, as you stir the solution continuously with the stirring rod. Continue adding the acid until the color *just* disappears when the solution is stirred. Test the solution by transferring a drop to a strip of universal indicator paper in the watch glass. Match the color change on the strip with the color standard, then record the pH of the mixed solution in the data table.

4. Place the evaporating dish and solution on an evaporating setup as shown in figure 34.1. Add a few more drops of hydrochloric acid to the solution and stir with a stirring rod before you begin heating. This will ensure that neutralization is complete and the excess acid will be vaporized during the evaporation process. Evaporate the solution to dryness. Allow the dish to cool, then examine the residue. Record your observations in Data Table 34.1.

5. Dissolve the residue in 2 mL of distilled water. Test the dissolved residue with universal indicator paper and record the pH in Data Table 34.1.

Part B: Preparation of Potassium Sulfate

Repeat the entire part A procedure, this time using solutions of potassium hydroxide and sulfuric acid. Record your findings for this reaction in Data Table 34.2.

Results

1. What ions are present in solutions of (a) sodium hydroxide, (b) hydrochloric acid, and (c) a mixture of hydrochloric acid and sodium hydroxide after they react?

2. Describe experimental evidence to support your answers for each part of question 1.

3. What ions are present in solutions of (a) potassium hydroxide, (b) sulfuric acid, and (c) a mixture of sulfuric acid and potassium hydroxide after they react?

4. Describe experimental evidence to support your answers for each part of question 3.

5. Write balanced equations for the two neutralization reactions carried out in this experiment.

6. Was the purpose of this lab accomplished? Why or why not? (Your answer to this question should show thoughtful analysis and careful, thorough thinking.)

Invitation to Inquiry

Salinity is a measure of the amount of salts dissolved in 1 kg of solution. For example, if 1 kg of seawater were evaporated, 965 g of water would leave and 35 g of salt would remain. This is a salinity of 35 parts per thousand, or 35‰. Measuring, then evaporating the water from a kilogram sample of water would be time consuming. To save time, design a way to measure salinity from conductivity. Construct a conductivity table for salt solutions of known concentrations, then use the table to check the salinity of a sample of sea water. You could also consider measuring the density of seawater by constructing a calibrated hydrometer.

Data Table 34.1   Production of Sodium Chloride

pH of hydrochloric acid solution	_____
pH of sodium hydroxide solution	_____
pH of hydrochloric acid and sodium hydroxide mixture	_____
Observations of residue	
pH of residue solution	_____

Data Table 34.2   Production of Potassium Sulfate

pH of sulfuric acid solution	_____
pH of potassium hydroxide solution	_____
pH of sulfuric acid and potassium hydroxide mixture	_____
Observations of residue	
pH of residue solution	_____

Experiment 35: Identifying Salts

Introduction

A **salt** is any ionic compound except those with hydroxide (OH⁻) or oxide (O⁻²) ions. Table salt, which is sodium chloride (NaCl), is but one example from the large group of ionic compounds known as salts. Simple salts consist of a metallic ion (such as Na⁺) in an ionic crystalline structure with nonmetallic ions (such as Cl⁻). When a salt is dissolved in water the ion crystal structure is separated into a solution of metallic and nonmetallic ions.

The name of a salt provides a clue about the ions present in a solution. Lithium chloride (LiCl) becomes Li⁺ and Cl⁻ ions when dissolved in water. Likewise, calcium iodide (CaI₂) becomes a solution of Ca⁺² and I⁻ ions, and iron(II) carbonate [Fe₂(CO₃)₃] becomes a solution of Fe⁺³ and CO₃⁻² ions. As the water of a salt solution is evaporated, the salt ions again form an ionic crystal structure. Knowing the ions present in a given salt solution will therefore identify the salt in the solution. In this experiment, you will use a flame test to identify metallic ions present in a salt solution. Each metal ion gives a characteristic color to a burner flame. The nonmetallic ions will be identified by chemical tests.

CAUTION: Acids will damage human tissue and other materials. Silver nitrate will stain the skin. Some of the chemicals are poisonous when swallowed. Be sure to wash thoroughly if any chemicals are spilled and inform your instructor. Wash your hands thoroughly after the experiment.

Materials

Beakers, burner, medicine dropper, cobalt glass squares, platinum or nichrome wire, forceps, test tubes and rack, graduated cylinder, pipette (or thin glass tube). **Flame test solutions:** sodium nitrate solution (1.0 M: 85 g/L of solution), lithium nitrate (1.0 M: 123 g/L of solution), strontium nitrate (1.0 M: 284 g/L of solution), calcium nitrate (1.0 M: 163 g/L of solution), barium nitrate (1.0 M: 261 g/L of solution), potassium nitrate (1.0 M: 101 g/L of solution), dilute hydrochloric acid (1:4), and distilled water. **Chemical test solutions:** silver nitrate solution (0.1 M: 17.3 g/L of solution), barium chloride solution (0.1 M: 24.4 g/L of solution), calcium carbonate solution (saturated), iron(II) sulfate solution (saturated), concentrated sulfuric acid, sodium chloride solution (0.1 M: 5.9 g/L of solution), and potassium sulfate solution (0.1 M: 15.8 g/L of solution).

275

Procedure

Part A: Flame Tests

1. Obtain seven test tubes and a test tube rack. Wash the test tubes and rinse them thoroughly with distilled water. Cleanliness is very important throughout this experiment.

2. Pour about 25 mL of dilute hydrochloric acid into one test tube and label it in a rack. Pour about 5 mL of each of the six nitrate solutions into test tubes and label each in the rack. Set up a burner and adjust it for an inner blue cone and no yellow color.

3. Clean a platinum or nichrome wire by dipping it into dilute hydrochloric acid, then holding it in the pale blue part of the flame. Repeat the cleaning procedure until the wire gives no color to the flame. The flame test you are about to do is a very sensitive test and even trace contamination of sodium from your fingers will give the characteristic color of the sodium ion (a yellow coloration).

4. Dip the clean wire into one of the solutions and place just the tip of the wire into the light blue burner flame. Record your observations in Data Table 35.1.

5. Clean the wire, then repeat the flame test with each of the solutions. Record the results obtained for each in Data Table 35.1. Use two thicknesses of cobalt glass to view the potassium and sodium flames. The sodium ion gives a color that masks the color given by the potassium ion. The cobalt glass filters out the color produced by any sodium ions.

Part B: Chemical Tests

1. Obtain four clean test tubes and a test tube rack. Pour about 5 mL of each solution of sodium chloride, potassium sulfate, calcium carbonate, and one of the nitrate solutions from the flame test into separate test tubes. Label each in the test tube rack.

2. **Chloride ion test**: To 5 mL of sodium chloride solution, add 5 drops of silver nitrate solution. Mix the solutions well, then describe the results in Data Table 35.2.

3. **Sulfate ion test**: To 5 mL of potassium sulfate solution, add 10 drops of barium chloride solution. Mix well, then describe the results in Data Table 35.2.

4. **Carbonate test**: To 5 mL of calcium carbonate solution, add several drops of dilute hydrochloric acid. Describe the results in Data Table 35.2.

5. **Nitrate ion test**: To 5 mL of a nitrate solution, add 10 mL of iron(II) sulfate solution. Use a long, thin pipette to slowly add 2 mL of concentrated sulfuric acid so it runs down the inside of the tube without mixing. Observe the interface between the acid and the solution, recording the results in Data Table 35.2.

Part C: Unknown Solutions

1. Wash test tubes of all salt solutions and rinse thoroughly with distilled water.

2. Your instructor will supply unknown solutions, each containing a single metallic and a single nonmetallic ion. Using the procedures and findings from part A and part B, identify the ions in the unknown solutions. Record your findings in Data Table 35.3, describing your test results as very positive, positive, trace, negative, or unsure.

Invitation to Inquiry

Foods naturally contain enzymes, biochemical compounds that originate in plants and animals. Cooking changes enzymes, and the purpose of blanching, or steaming food shortly before freezing is intended to destroy enzymes. One of the enzymes in foods is a catalysts that will speed the decomposition of hydrogen. You can design a home experiment to use hydrogen peroxide (3% solution) to test fresh, blanched, and cooked crushed food (or juices from the foods) for the catalyst enzymes. If you can find a way to quantify the measurements, perhaps you can come up with specific recommendations about how hot the food should be heated, and for how long.

Temperature is one of the more important factors that influence the rate of a chemical reaction. You can use a "light stick" or "light tube" to study how temperature can influence a chemical reaction. Light sticks and tubes are devices that glow in the dark and have become very popular on July 4th, Halloween, and other times why people might be outside after sunset. They work from a chemical reaction that is similar to the chemical reaction that produces light in a firefly. Design a home experiment that uses light sticks to find out the effect of temperature on the brightness of light and how long the device will continue providing light. Perhaps you will be able to show by experimental evidence that use at a particular temperature produces the most light for the longest period of time.

Metallic ion	Results
Sodium	
Lithium	
Strontium	
Calcium	
Barium	
Potassium	

Data Table 35.1   Flame Tests for Metallic Ions

Data Table 35.2	Chemical Tests for Nonmetallic Ions
Nonmetallic ion	Results
Chloride	
Sulfate	
Carbonate	
Nitrate	

Data Table 35.3		Unknown Solutions	
Unknown	Metallic Ion Tests	Nonmetallic Ion Tests	Compound
1	Sodium: Lithium: Strontium: Calcium: Barium: Potassium:	Chloride: Sulfate: Carbonate: Nitrate:	Name: _____ Formula: _____
2	Sodium: Lithium: Strontium: Calcium: Barium: Potassium:	Chloride: Sulfate: Carbonate: Nitrate:	Name: _____ Formula: _____
3	Sodium: Lithium: Strontium: Calcium: Barium: Potassium:	Chloride: Sulfate: Carbonate: Nitrate:	Name: _____ Formula: _____

Experiment 36: Solubility Curves

Introduction

A **solution** is a homogeneous mixture of ions or molecules of two or more substances. A solution is made by **dissolving**; that is, mixing the different components that make up a solution. When sugar is dissolved in water, for example, the molecules of sugar become evenly dispersed throughout the molecules of water. When evenly mixed, a uniform taste of sweetness is observed in any part of the sugar solution.

There is a limit to how much solid can be dissolved in a certain amount of liquid at a given temperature. When the limit is reached, more added solid does not dissolve but accumulates at the bottom of the solution. At its limit the solution is described as a **saturated solution**. A saturated solution cannot dissolve any more solute at a given temperature. The process is not a static one, however, as dissolving and crystallization continue in a saturated solution. But now the rate of dissolving equals the rate of solute coming out of solution.

The amount of a given solute that will dissolve in a certain amount of water varies with the particular solute as well as with the temperature. The **solubility** of a solute is the concentration that is reached in a saturated solution at a particular temperature. For example, the solubility of a given salt in a specific amount of water varies with the temperature. A graph of solubility and temperature is called a **solubility curve**. In this experiment you will find the solubility curve for potassium dichromate (or some other salt as specified by your instructor) in a group experiment. Teams of students will find solubility values at different temperatures. A solubility curve will be drawn from the results of the whole laboratory group.

Materials

Ring stand and ring, burner, beakers, evaporating dish, wire screen, balance, graduated cylinder, distilled water, crushed ice, potassium dichromate, thermometer, glass stirring rod, spatula, laboratory forceps.

Procedure

Part A: Preparation of Saturated Solution

1. This is a group experiment. Teams of students will prepare saturated solutions of potassium dichromate at temperatures of about 85, 75, 65, 55, 45, 35, 25, and 20° C. The range and temperature intervals may be adjusted according to the size of the laboratory class, but each team

will be assigned a particular temperature to investigate. The results of each team will be pooled to form a class solubility curve.

2. Determine which team will be responsible for investigating the solubility of potassium dichromate at which temperature. Record the assigned temperature you will investigate in Data Table 36.1.

3. Obtain a clean test tube and a beaker that will hold the test tube. Prepare a water bath by adding sufficient water to the beaker to cover the bottom half of the test tube when it is in the beaker. Using hot water or crushed ice as necessary, adjust the temperature of the water to your assigned value.

4. Add 25 mL of distilled water to the test tube. Place the test tube in the water bath, stirring the water in the test tube until an equilibrium is reached near the assigned value. Record the actual temperature reached in Data Table 36.1.

5. With the test tube still in the water bath, use a spatula to carefully add small amounts of potassium dichromate, stirring the solution with a glass rod. Continue adding small amounts of potassium dichromate and stirring until no more will dissolve. You should now have a saturated solution at or nearly at the assigned temperature.

Part B: Measuring Solubility

1. Set up the apparatus as shown in figure 36.1, but without the evaporating dish. Start heating the water in the beaker. Accurately measure the mass of the clean, dry evaporating dish and record this value in Data Table 36.1.

2. Pour about 20 mL of your saturated solution into the evaporating dish, making sure that no undissolved crystals are added to the dish. Accurately measure the mass of the dish and solution, recording the value in the data table.

3. Place the evaporating dish on the beaker of gently boiling water. Continue heating until the solution in the dish is evaporated to dryness.

4. Place the evaporating dish on a wire screen on a ring stand. *Gently* heat the dish with a small burner flame to remove any remaining traces of water. Avoid rapid or strong heating which could cause splattering or decomposition of the salt.

5. Cool, then accurately measure the mass of the dish and dry salt. Record the mass, then do the necessary calculations to complete Data Table 36.1.

6. Plot the laboratory class results with the temperature on the *x*-axis and the solubility on the *y*-axis.

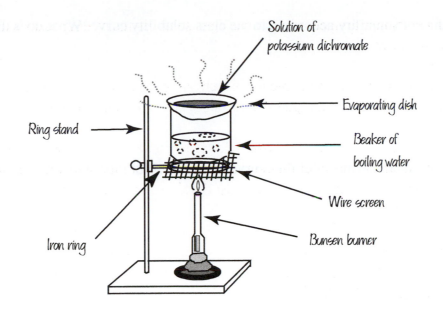

Solution of potassium dichromate

Evaporating dish

Ring stand

Beaker of boiling water

Wire screen

Bunsen burner

Iron ring

Figure 36.1

Results

1. Describe the solubility curve produced from the overall class data.

2. How is this solubility curve similar to or different from the solubility curve of other salts? Give specific comparisons.

283

3. Describe the range of solubility according to the class solubility curve. What does this mean?

4. Describe how you could use the solubility curve to predict solubility at *any* temperature.

5. Was the purpose of this lab accomplished? Why or why not? (Your answer to this question should show thoughtful analysis and careful, thorough thinking.)

Invitation to Inquiry

What factor is most important in determining the rate of the chemical reaction? For example, is temperature, surface area, agitation, or concentration of reactants most important in determining how fast a reaction takes place? Or perhaps they are all equally important? Design your own experiment to answer these questions. Use Alka-Seltzer® tablets that you will dissolve in water, and design an experiment to find the best way to increase the dissolving rate.

How can you measure the dissolving rate of Alka-Seltzer® in water? Consider collecting the gas given off by water displacement, using a stop watch to determine rate. See laboratory experiments in this manual concerned with oxygen and/or hydrogen for an experimental setup on collecting a gas by water displacement.

Of those listed—temperature, surface area, agitation, or concentration—which of the factors tested were important in determining the rate of the reaction?

Data Table 36.1   Solubility at Specific Temperature

1. Assigned temperature -- _____ °C

2. Actual temperature investigated ------------------------------------ _____ °C

3. Mass of clean, dry evaporating dish ---------------------------------- _____ g

4. Mass of dish and solution -- _____ g

5. Mass of solution (row 4 – row 3) -------------------------------------- _____ g

6. Mass of dish and dry salt -- _____ g

7. Mass of dry salt (row 6 – row 3) -------------------------------------- _____ g

8. Mass of water in solution (row 5 – row 7) -------------------------- _____ g

9. Solubility: Mass of salt/100 g water (row 7 ÷ row 8 × 100) ---- _____ g

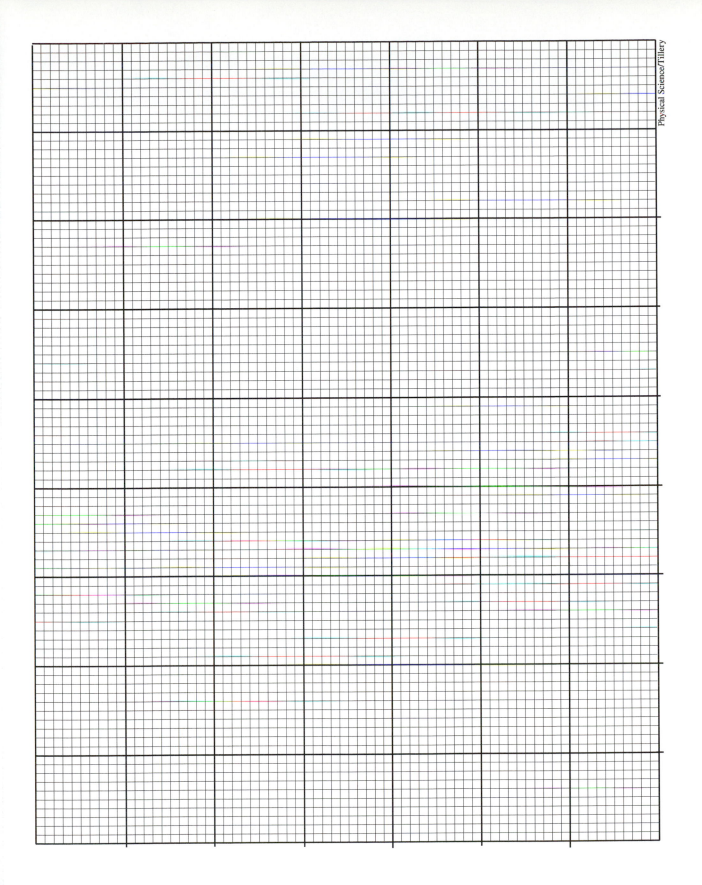

Experiment 37: Natural Water

Introduction

Water is sometimes referred to as the "universal solvent" because so many gases, liquids, and solids readily dissolve in it. In addition, the solubility of carbon dioxide in water produces an acid (H_2CO_3) that contributes to the dissolution of normally insoluble carbonate, phosphate, and sulfite minerals. Thus natural water can contain a significant amount of dissolved mineral matter as well as suspended solids. This occurs naturally from weathering and erosion of the earth's surface by rainwater and by erosion of stream beds.

Natural water that contains dissolved mineral salts in the form of calcium ions or magnesium ions is called **hard water**. It is called hard water because the metal ions react with soap, making it hard to make soap lather. Instead of producing suds, soap in hard water produces an insoluble, sticky precipitate. This requires a greater amount of soap to produce suds since all the metal ions must be precipitated out of solution before the soap will lather. The precipitate also produces a bathtub ring, collects on clothes, and results in other undesirable effects. When boiled, hard water can produce a solid deposit that restricts water flow in water heaters and pipes.

Water hardness is usually measured in parts per million (ppm), with hard water identified as any concentration of calcium or magnesium ions greater than about 75 ppm. In this experiment you will learn a method of analyzing water hardness as well as analyzing the type of hardness present in your water supply.

Materials

Distilled water, 3% alum solution, aqueous ammonium hydroxide (6 M), beakers, dropper, glass stirring rod, filtering funnel, ring stand and ring, glass wool, clean sand, test tubes and rack, wire screen, burner, watch glass, graduated cylinder, calcium chloride solution (0.001 M: 0.1 g/L of solution), standard soap solution, muddy water, tap or well water.

Procedure

Part A: Suspended Solids

1. Natural water usually contains suspended solids that are removed by (a) natural settling, (b) addition of a gelatinous precipitate, and (c) filtering through sand. Compare these methods by obtaining four beakers, each with 100 mL of muddy water.

2. Allow beaker #1 of muddy water to stand undisturbed as you complete the rest of this section. This will provide a comparison of natural settling to the other methods of removing suspended solids.

3. Make a sand filter by placing a loose plug of glass wool in a funnel, then covering it with several cm of clean sand. Pour beaker #2 of muddy water through the filter and collect the filtrate in a clean beaker.

4. In beakers #3 and #4, produce the gelatinous precipitate of aluminum hydroxide by adding to each 10 mL of alum solution, then 4 drops of ammonium hydroxide while stirring. Allow both suspensions to settle.

5. Prepare a second sand filter as in procedure step 3, then filter one of the settled beakers of aluminum hydroxide suspension from procedure step 4. Collect this filtrate in a clean beaker and save it for part C of this experiment.

6. Compare the appearance of the muddy water from each procedure step in Data Table 37.1.

Part B: A Test for Hard Water

1. The calcium and magnesium ions of hard water react with soap, forming an insoluble precipitate. Soap will form suds only after the ions have been removed, so the amount of soap needed to produce suds is an indication of the amount of water hardness.

2. Measure 5 mL of the calcium chloride solution and pour it into a test tube. This solution has been prepared to have a water hardness of 100 ppm.

3. Add one drop of soap solution to the calcium chloride solution and insert a clean stopper. Shake the test tube vigorously, then check the surface for soap suds. If none are visible, or if visible suds do not persist for at least one minute, add a second drop and shake the test tube vigorously again. Continue adding one drop at a time and shaking, keeping track of the number of drops added until a layer of suds persists for at least a minute. Record the number of drops required for permanent suds in Data Table 37.2.

4. Repeat procedure step 3 with a clean test tube and 5 mL of distilled water. A certain number of soap drops will be required to make a permanent suds layer because of the mechanism by which suds are produced. In Data Table 37.2, record the number of soap drops required for this mechanism to occur in distilled water.

5. Complete the calculations required in Data Table 37.2 to find the hardness equivalent to one drop of the soap solution.

Part C: Water Hardness

1. Repeat procedure step B-3 with 5 mL of the flocculated and filtered water saved from part A of this experiment. Record the number of drops of soap required in Data Table 37.3 and calculate the hardness of the water.

2. Repeat the soap-drop test with 5 mL of ordinary cold tap or well water. Record the number of soap drops required in Data Table 37.3 and calculate the water hardness.

3. Water hardness caused by calcium or magnesium bicarbonate can be removed by boiling, so it is called temporary hardness. The hardness remaining after boiling is permanent hardness. Temporary hardness is therefore the difference between total hardness and permanent hardness.

4. Pour 25 mL of cold tap water or well water into a small beaker. Place the beaker on a wire screen on a ring stand, with a clean watch glass on top of the beaker to prevent water loss by splattering. Gently heat the water to boiling, continuing a slow boil for 5 or 6 minutes.

5. Cool the water to room temperature, then pour the water into a graduated cylinder. Add distilled water as necessary to bring the volume back to 25 mL. Pour 5 mL into a test tube. Use the soap-drop method to determine the permanent hardness. Calculate the temporary hardness. Record all data in Data Table 37.3.

Results

1. Compare the appearance of muddy water after natural settling and the other means of removing suspended solids. Which method produced the best apparent purity? Is there any impurity that the best method does not remove?

2. Explain how a soap solution was standardized to measure water hardness.

3. What is hard water?

4. Explain how temporary and permanent hardness can be measured.

5. What is the white solid that forms on water outlets in hard water areas? How could you test the substance to confirm your answer?

6. Was the purpose of this lab accomplished? Why or why not? (Your answer to this question should show thoughtful analysis and careful, thorough thinking.)

Invitation to Inquiry

Washing soda, borax, or trisodium phosphate are often added to laundry products to soften the wash water. Add a small amount of each to 5 mL of tap water, then measure the hardness with the soap-drop method. Is one of the softening chemicals more effective than the others?

Is there a practical way to obtain pure water from a hard water source? Design, use, and evaluate an apparatus for purifying water—by boiling, freezing, filtering, precipitation of dissolved minerals—or by any means you can imagine. Evaluate your technique in terms of usefulness and effectiveness.

Data Table 37.1	Removing Suspended Solids
Action	Result
1. Initial appearance of muddy water	
2. After natural settling	
3. After sand filtration only	
4. After aluminum hydroxide flocculation and settling	
5. After aluminum hydroxide flocculation, settling, and sand filtering.	

Data Table 37.2   Hardness Equivalent of Soap Solution

1. Drops of soap solution required for calcium chloride solution (100 ppm)	_____ drops
2. Drops of soap solution for distilled water -------------------------	_____ drops
3. Subtract sudsing mechanism (row 1 – row 2) --------------------	_____ drops
4. Hardness equivalent (100 ppm ÷ row 3) -------------------------	_____ ppm/drop

Data Table 37.3   Permanent, Temporary, and Total Hardness of Tap Water

1. Drops of soap solution required for filtered water --------------- _____ drops

2. Water hardness (row 1 × ppm/drop from Data Table 37.2) ------ _____ ppm

3. Drops of soap solution required for tap water -------------------- _____ drops

4. Total water hardness of tap water -------------------------------- _____ ppm

5. Drops of soap solution required after boiling --------------------- _____ drops

6. Permanent hardness of tap water --------------------------------- _____ ppm

7. Temporary hardness of tap water (row 4 – row 6) --------------- _____ ppm

Experiment 38: Measurement of pH

Introduction

Acids and bases are classes of chemical compounds and each class has certain properties in common. Acids have characteristic properties of reacting with active metals to release hydrogen gas, neutralizing bases to form a salt, and changing the color of certain substances such as litmus. Bases, on the other hand, have characteristic properties of converting plant and animal tissues to soluble substances, neutralizing acids to form a salt, and reversing the color changes that were caused by acids.

Some chemical compounds are acidic, such as vinegar, and other compounds are basic, such as lye; but the terms acidic and basic are very general and inexact. The **pH scale** is a concise and quantitative way of expressing the strength of an acid or a base. On this scale a neutral solution (neither acidic nor basic properties) has a pH of 7.0. Acid solutions have pH values below 7 and smaller numbers mean greater acidic properties. Basic solutions have pH values above 7 and larger numbers mean greater basic properties. The pH scale is logarithmic, so a pH of 2 is 10 times more acidic than a pH of 3. Likewise, a pH of 8 is 10 times 10, or 100 times more basic than a pH of 6.

Many plant extracts and synthetic dyes change colors when mixed with acids or bases. Such substances that change color in the presence of acids or bases can be used as an **acid-base indicator**. Litmus, for example, is an acid-base indicator made from a dye extracted from certain species of lichens. The dye is applied to paper strips, which turn red in acidic solutions and blue in basic solutions. This is only a "ball park" indicator, however, since it only indicates on which side of pH 7 a solution lies. Other indicators are available that can be used to estimate the pH to about half a pH unit — for example, a pH of 3.5. There are also electrical instruments that measure pH by measuring the conductivity of a solution. In this experiment you will determine the characteristics of several commercial indicators, then test unknown solutions for acid and base properties.

Materials

Solutions of: Hydrochloric acid (0.1 M: 8.1 mL concentrated/L of solution), sulfuric acid (0.1 M: 5.6 mL concentrated/L of solution), acetic acid (0.1 M: 5.6 mL concentrated/L of solution), sodium hydroxide (0.1 m: 4 g/L of solution), barium hydroxide (17 g/L of solution), household ammonia (diluted with 6 volumes water). **Indicators**: Red litmus paper, blue litmus paper, bromthymol blue, methyl orange, methyl red, phenolphthalein, and universal indicator paper (with color-coded pH scale). Watch glass, glass stirring rod, test tubes and rack. Various unknown solutions (different acid or base solutions of different concentrations).

CAUTION: Acids and bases can damage skin and other materials. If spilled on your person or clothing, flush with running water for several minutes and inform the instructor. Dilute any tabletop spills with water, then use a sponge to wipe up the spill. Be sure to rinse the sponge with plenty of water.

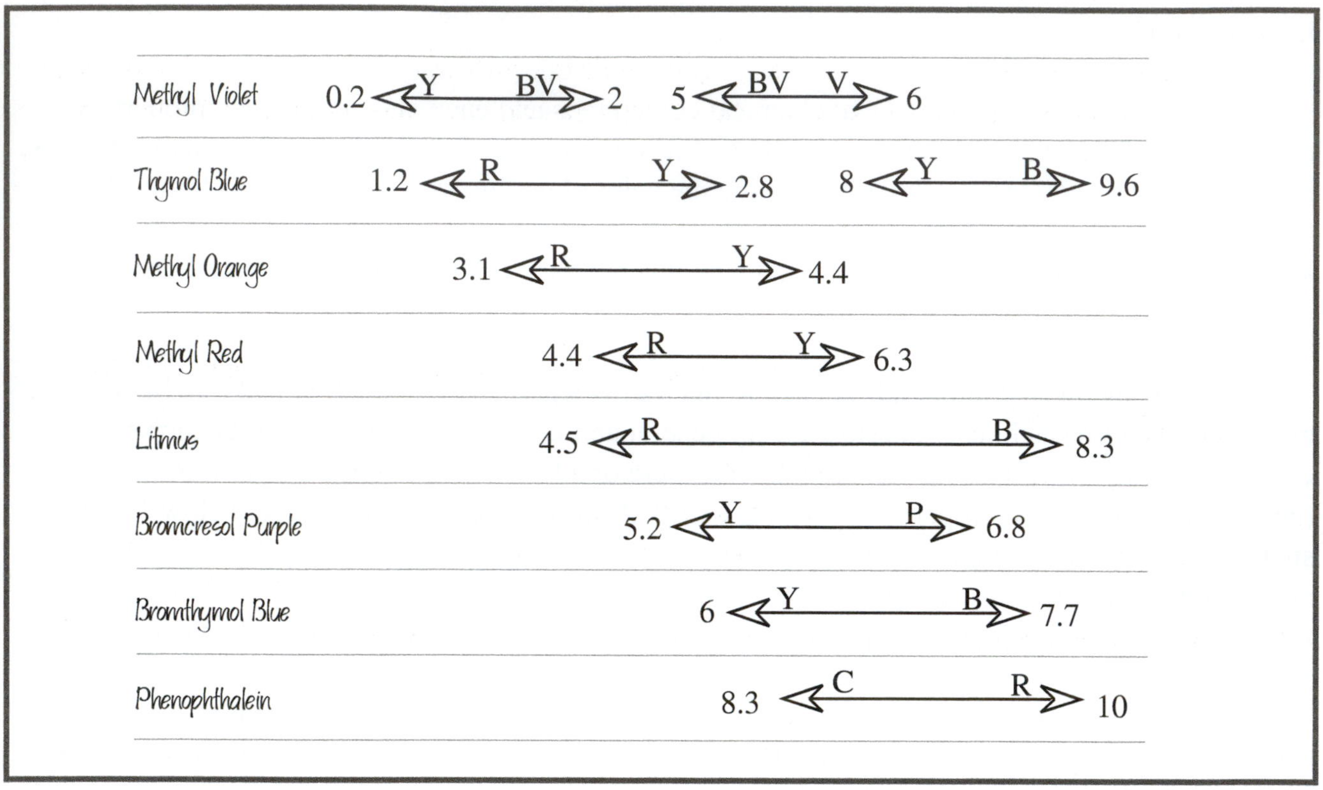

Methyl Violet 0.2 ⟵Y BV⟶ 2   5 ⟵BV V⟶ 6

Thymol Blue 1.2 ⟵R Y⟶ 2.8   8 ⟵Y B⟶ 9.6

Methyl Orange 3.1 ⟵R Y⟶ 4.4

Methyl Red 4.4 ⟵R Y⟶ 6.3

Litmus 4.5 ⟵R B⟶ 8.3

Bromcresol Purple 5.2 ⟵Y P⟶ 6.8

Bromthymol Blue 6 ⟵Y B⟶ 7.7

Phenophthalein 8.3 ⟵C R⟶ 10

Figure 38.1 pH Ranges and Color Changes for Selected Indicators

Procedure

1. Obtain five clean test tubes and a test tube rack. Pour 5 mL of hydrochloric acid into each. Test the hydrochloric acid with litmus by placing a strip of red litmus paper and a strip of blue litmus paper in a clean watch glass. Dip a clean, dry stirring rod into the solution in one of the test tubes, then transfer a drop to each litmus paper. Record your observations in Data Table 38.1. Save the litmus paper and watch glass for further testing.

2. Add two drops of each indicator solution to a test tube of hydrochloric acid, recording your observations in Data Table 38.1. Include pH values or ranges where possible.

3. Place a strip of universal indicator paper on the watch glass. Use the stirring rod to transfer a drop of hydrochloric acid to the indicator paper. The universal indicator paper container will have a color standard chart with a corresponding pH. Record the color and pH number in Data Table 38.1. If a color falls between two of the standards, estimate the pH to a fraction between the corresponding pH values.

4. Repeat procedure steps 1 through 3 using sulfuric acid, acetic acid, sodium hydroxide, barium hydroxide, and dilute household ammonia solutions. Record the results for the acid solutions in Data Table 38.1 and the results for the base solutions in Data Table 38.2. When all of the solutions have been tested, take a few minutes to analyze your findings. Compare the pH ranges of the indicators in table 38.1 with the pH as shown by the universal indicator.

5. Obtain an unknown solution and use the different indicator tests to determine if the solution is an acid or a base and to determine the pH of the solution. Record your test results and findings in Data Table 38.3.

Results

1. A popular noncarbonated beverage turns thymol blue to yellow and methyl orange to red. What is the approximate pH of this soft drink?

2. What is the approximate pH of a beer that turns methyl orange to a yellow color?

3. How would the color changes of a mixture of methyl orange, methyl red, bromthymol blue, and phenolphthalein compare with those of the universal indicator?

4. Would you use a solution of phenolphthalein to find the strength of an acid? Explain.

5. The flowering shrub hydrangea produces blue flowers in a certain soil, but produces pink flowers if lime is added to the soil. Propose an explanation for this observation. Explain how you could do tests on the soil to check your explanation.

6. Was the purpose of this lab accomplished? Why or why not? (Your answer to this question should show thoughtful analysis and careful, thorough thinking.)

Invitation to Inquiry

There are many common things that can be used as acid and base indicators. For example, from foods you could try boiled purple cabbage juice, grape juice, blackberry juice, ordinary tea, and others, and from the office supply store certain construction papers might show color changes. To test foods, try soaking strips of filter paper in juices or solutions, then allowing it to dry completely. Materials such as construction paper can be tested directly. Plan tests to find what color each indicator will appear when exposed to solutions of acids or bases, and test other materials, too. Show how your collections of indicators will identify the pH from a wide range of possibilities.

Think of some way to measure the "strength" of an acid or a base without using an indicator. Taking proper precautions, do experiments to compare several methods of measuring strength.

Indicator	Hydrochloric Acid	Sulfuric Acid	Acetic Acid
Red Litmus			
Blue Litmus			
Bromthymol Blue			
Methyl Orange			
Methyl Red			
Phenolphthalein			
Universal Indicator Color			
Universal Indicator pH			

Data Table 38.1   Results of Indicator Tests for Acid Solutions

Data Table 38.2	Results of Indicator Tests for Base Solutions		
Indicator	Sodium Hydroxide	Barium Hydroxide	Dilute Ammonia
Red Litmus			
Blue Litmus			
Bromthymol Blue			
Methyl Orange			
Methyl Red			
Phenolphthalein			
Universal Indicator Color			
Universal Indicator pH			

Data Table 38.3	Test Results for Unknown Solution
Indicator	Results
Red Litmus	
Blue Litmus	
Bromthymol Blue	
Methyl Orange	
Methyl Red	
Phenolphthalein	
Universal Indicator Color	

pH of unknown solution: _____

Is the unknown solution an acid or a base? _____

Experiment 39: Amount of Water Vapor in the Air

Introduction

The amount of water vapor in the air is referred to generally as **humidity**. A measurement of the amount of water vapor in the air at a particular time is called the **absolute humidity**. At room temperature, for example, humid air might contain 15 grams of water vapor in each cubic meter of air. At the same temperature air of low humidity might have an absolute humidity of only 2 grams per cubic meter. Absolute humidity can range from near zero up to a maximum that is determined by the temperature at a particular time, as shown in figure 39.1. Since the temperature of the water vapor present in the air is the same as the temperature of the air, the maximum absolute humidity is usually said to be determined by the air temperature. What this really means is that the maximum absolute humidity is determined by the temperature of the water vapor; that is, the average kinetic energy of the water vapor.

The relationship between the *actual* absolute humidity at a particular temperature and the *maximum* absolute humidity that can occur at that temperature is called the **relative humidity**. Relative humidity is a ratio between (1) the amount of water vapor in the air, and (2) the amount of water vapor needed to saturate the air at that temperature. The relationship is

$$\frac{\text{actual absolute humidity at present temperature}}{\text{maximum absolute humidity at present temperature}} \times 100\% = \text{relative humidity}$$

For example, suppose a measurement of the water vapor in the air at 10° C (50° F) finds an absolute humidity of 5.0 g/m³. According to figure 39.1, the maximum amount of water vapor that can be in the air when the temperature is 10° C is about 10 g/m³. The relative humidity is then

$$\frac{5.0 \text{ g/m}^3}{10 \text{ g/m}^3} \times 100\% = 50\%$$

If the absolute humidity were 10 g/m³ at this temperature, then the air would have all the water vapor it could hold and the relative humidity would be 100%. A relative humidity of 100% means that the air is saturated at the present temperature.

Procedure

Part A: Maximum Amount of Water Vapor

1. Measure the present air temperature in your laboratory room and record it in Data Table 39.1. Use the graph of maximum absolute humidity in figure 39.1 to estimate the *maximum* amount of water vapor a cubic meter of air can hold at this temperature. Since one gram of water has an approximate volume of one milliliter, find the maximum amount of water in liters per cubic meter that can be in the room at the present temperature. Record this maximum, in grams/cubic meter and in liters/cubic meter, in Data Table 39.1.

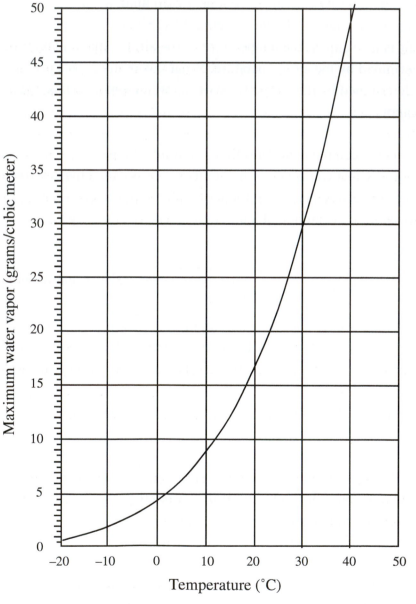

Figure 39.1

2. Measure the length, width, and height of the laboratory room. Record these measurements in Data Table 39.1, then calculate the volume of the room in cubic meters. Record all data and calculations in the data table.

3. Calculate the maximum amount of water vapor the air in the laboratory room can hold at the present temperature. This is found by multiplying the volume of the room (in cubic meters) by the maximum amount of water vapor that could be in the room at the present temperature (in liters per cubic meter). Record all data and calculations in Data Table 39.1.

Part B: Actual Amount of Water Vapor

Evaporation occurs at a rate that is inversely proportional to the relative humidity, ranging from a maximum rate when the air is driest to no net evaporation when the air is saturated. Since evaporation is a cooling process, it is possible to use a thermometer to measure humidity. An instrument called a **psychrometer** has two thermometers, one of which has a damp cloth wick around its bulb end. As air moves past the two thermometer bulbs, the ordinary thermometer (the dry bulb) will measure the present air temperature. Water will evaporate from the wet wick (the wet bulb) until an equilibrium is reached between water vapor leaving the wick and water vapor returning to the wick from the air. Since evaporation lowers the temperature, the depression of the temperature of the wet-bulb thermometer is an indirect measure of the water vapor present in the air. The relative humidity can be determined by obtaining the dry- and wet-bulb temperature readings and referring to a relative humidity chart such as the one found inside the back cover. If the humidity is 100%, no net evaporation will take place from the wet bulb, and both wet- and dry-bulb temperatures will be the same. The lower the humidity, the greater the difference in the temperature reading of the two thermometers.

Relative humidity is a ratio between the actual absolute humidity at a given temperature and the maximum absolute humidity that can occur at that temperature. Knowing the maximum absolute humidity and the relative humidity, you can find the amount of water vapor in the air at the present temperature.

1. Wet the cotton wick on the wet bulb of a sling psychrometer. Whirl the thermometers in the air until the wet-bulb thermometer registers its lowest reading. Record the wet-bulb and dry-bulb temperatures, then use this data to find the relative humidity from the relative humidity chart inside the back cover. Record all data and calculations in the data table.

2. Multiply the humidity as a fraction times the maximum amount of water vapor that could be in your laboratory room at the present temperature. Record the amount of water vapor present in the room in the data table.

Results

1. What is the actual amount of water vapor present in the laboratory room air at the present temperature (in g/m^3)? What is the maximum amount of water vapor that *could* be present?

2. Can the absolute humidity of the air in the room be increased? Explain.

3. Can the relative humidity of the air in the room be increased without adding more water vapor to the air? Explain.

4. Suppose the room air has all the water vapor it will hold at 25° C, and the air is cooled to 15° C. Considering the area of the floor from your measurements, how deep a layer of water will condense from the air?

5. Was the purpose of this lab accomplished? Why or why not? (Your answer to this question should show thoughtful analysis and careful, thorough thinking.)

Invitation to Inquiry

Cobalt chloride is often used to test for the presence of water since it undergoes a reversible color change when exposed to moisture or humidity. For example, cobalt chloride is sometimes included with silica gel pellets to indicate when the gel has absorbed moisture. Experiment with cobalt chloride dried on filter paper strips. Find out if the color change is sensitive enough to water vapor and temperature to be used as an indicator of the relative humidity.

Data Table 39.1	Amount of Water Vapor in the Laboratory Room	
1. Present temperature of laboratory room air	_____	°C
2. Maximum absolute humidity at present temperature		
In grams per cubic meter	_____	g/m^3
In liters per cubic meter	_____	L/m^3
Room length	_____	m
Room width	_____	m
Room height	_____	m
3. Volume of laboratory room (length × width × height)	_____	m^3
4. Maximum amount of water vapor (row 2 × row 3)	_____	L
5. Dry-bulb reading	_____	°C
Wet-bulb reading	_____	°C
Difference in wet- and dry-bulb readings	_____	°C
Relative Humidity (from relative humidity chart)	_____	%
6. Amount of water vapor in room (row 5 as decimal × row 4)	_____	L

Experiment 40: Nuclear Radiation

Introduction

All the isotopes of all the elements with an atomic number greater than 83 (bismuth) are radioactive. **Radioactivity** is defined as the *spontaneous emission of particles or energy from an atomic nucleus* as it disintegrates. As a result of the disintegration the nucleus of an atom often undergoes a change of identity, becoming a simpler nucleus. The natural spontaneous disintegration or decomposition of a nucleus is also called **radioactive decay**. Although it is impossible to know when a given nucleus will undergo radioactive decay it is possible to deal with the rate of decay for a given radioactive material with precision.

An unstable nucleus decays into a more stable one by radioactive decay to become a more stable nucleus with less energy. There are five common types of radioactive decay and three of these involve alpha, beta, and gamma radiation.

• **Alpha emission.** Alpha (α) emission is the expulsion of an alpha particle from an unstable, disintegrating nucleus. The alpha particle, a helium nucleus, travels from 2 to 12 cm through the air, depending on the energy of emission from the source. An alpha particle is easily stopped by a sheet of paper close to the nucleus. As an example of alpha emission, consider the decay of a radon-222 nucleus,

$$^{222}_{86}\text{Rn} \rightarrow {}^{218}_{84}\text{Po} + {}^{4}_{2}\text{He}$$

The spent alpha particle eventually acquires two electrons and becomes an ordinary helium atom.

• **Beta emission.** Beta (β^-) emission is the expulsion of a different particle, a beta particle, from an unstable disintegrating nucleus. A beta particle is simply an electron ejected from the nucleus at a high speed. The emission of a beta particle *increases the number of protons* in a nucleus. It is as if a neutron changed to a proton by emitting an electron, or

$$^{1}_{0}\text{n} \rightarrow {}^{1}_{1}\text{p} + {}^{0}_{-1}\text{e}$$

Carbon-14 is a carbon isotope that decays by beta emission:

$$^{14}_{6}\text{C} \rightarrow {}^{14}_{7}\text{N} + {}^{0}_{-1}\text{e}$$

Note that the number of protons increased from six to seven, but the mass number remained the same. The mass number is unchanged because the mass of the expelled electron (beta particle) is negligible.

Beta particles are more penetrating than alpha particles and may travel several hundred cm through the air. They can be stopped by a thin layer of metal close to the emitting nucleus, such as a 1 cm thick piece of aluminum. A spent beta particle may eventually join an ion to become part of an atom, or it may remain a free electron.

• **Gamma emission.** Gamma (γ) emission is a high energy burst of electromagnetic radiation from an excited nucleus. It is a burst of light (photon) of wavelength much too short to be detected by the eye. Other types of radioactive decay, such as alpha or beta emission, sometimes leave the nucleus with an excess of energy, a condition called an *excited state*. As in the case of excited electrons, the nucleus returns to a lower energy state by emitting electromagnetic radiation. From a nucleus, this radiation is in the high-energy portion of the electromagnetic spectrum. Gamma is the most penetrating of the three common types of nuclear radiation. Like X rays, gamma rays can pass completely through a person but most gamma radiation can be stopped by a 5 cm thick piece of lead. As other types of electromagnetic radiation, gamma radiation is absorbed by and gives its energy to materials. Since the product nucleus changed from an excited state to a lower energy state, there is no change in the number of nucleons. For example radon-222 is an isotope that emits gamma radiation:

$$^{222}_{86}\text{Rn}^* \quad \rightarrow \quad ^{222}_{86}\text{Rn} \quad + \quad ^{0}_{0}\gamma$$

(* denotes excited state)

In this investigation you will become acquainted with some of the instrumentation of nuclear physics and examine the behavior of gamma radiation.

Procedure

Part A: Distance and Intensity of Nuclear Radiation

1. Turn on the Geiger counter and adjust it to the voltage range for the tube used. If the tube has a shield, it should be closed to absorb alpha and beta radiation. Thus you will be measuring gamma radiation only in this experiment.

2. Cover the tube with lead foil with a 3 mm hole and fix the tube in a vertical position as illustrated in figure 40.1. Note the hole should be centered at the middle of the tube. Read the *background* count in counts per minute (c/m) for 10 minutes and record it in Data Table 40.1. Make sure that all radioactive sources are at the far end of the room while measuring the background.

3. Place a radioactive source on a ring stand directly in line with the hole that is centered on the Geiger tube. Move the source close to the tube to obtain a high count-per-minute reading. Measure and record the distance and a corrected c/m reading in Data Table 40.1. The corrected c/m reading is obtained by subtracting the background count found in procedure step 2.

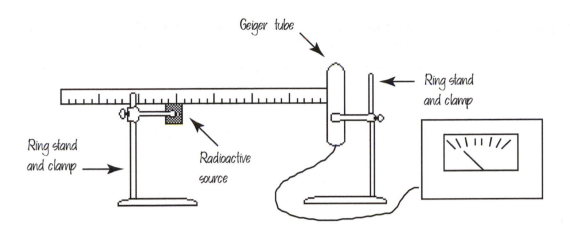

Figure 40.1

4. Move the source away from the Geiger counter tube in 1 cm intervals, recording the corrected c/m reading and the distance each time. Continue this process until there is no change for three successive intervals.

5. Plot the corrected c/m readings vs. distance on a full sheet of graph paper.

Part B: Absorption of Nuclear Radiation

1. Remove the lead foil from the Geiger tube, but leave the tube mounted in a vertical position. The tube measures alpha, beta, and gamma with the metal window open. It measures gamma only with the metal window closed. Adjust the voltage of the Geiger tube, then measure the background with the metal window closed. Record the background count in Data Table 40.2.

2. Place a radioactive source close enough to the tube for a high c/m reading. Record the reading, corrected for background, in Data Table 40.2.

3. Place a single sheet of lead foil between the radioactive source and the tube. The lead sheet should be as close to the radioactive source as possible. Record the corrected c/m reading for the attenuation property of a single sheet of lead.

4. Add additional sheets of lead foil close to the radioactive source, recording the corrected readings in Data Table 40.2.

5. Use a full sheet of graph paper to plot corrected c/m readings vs. number of lead sheets.

6. Your instructor might have other materials that can be studied concerning the passage of alpha, beta, and gamma rays.

Results

1. Describe how the behavior in count rate changes as a function of distance from a radioactive gamma source.

2. Generalize your description of c/m and distance by stating a quantitative relationship describing how the intensity of gamma radiation varies with distance from the source.

3. Would the generalization of question 2 apply to alpha and beta radiation as well as gamma? Explain why or why not.

4. What is the behavior in count rate as a function of thickness of a material for gamma rays?

5. How could you compensate if the actual count rate changes during a long counting period?

6. How does the penetrating ability of beta rays compare with that of gamma rays? How about alpha particles? Be quantitative in your answer.

7. Was the purpose of this lab accomplished? Why or why not? (Your answer to this question should show thoughtful analysis and careful, thorough thinking.)

Invitation to Inquiry

Ionizing radiation is understood to be potentially harmful if certain doses are exceeded. How much radiation do you acquire from the background where you live, from your lifestyle, and from medical procedures? Investigate radiation from cosmic sources, the sun, from television sets, from time spent in jet airplanes, and from dental or other X-ray machines. What are other sources of ionizing radiation in your community? How difficult is it to find relevant information and make recommendations? Does any agency monitor the amount of radiation that people receive? What are the problems and issues with such monitoring?

Data Table 40.1   Distance and Intensity of Nuclear Radiation

Background count _____ c/m

Distance (m)	Intensity (c/m)	Corrected Intensity (c/m)
_____	_____	_____
_____	_____	_____
_____	_____	_____
_____	_____	_____
_____	_____	_____
_____	_____	_____
_____	_____	_____
_____	_____	_____
_____	_____	_____

Data Table 40.2	Distance and Intensity of Nuclear Radiation	

Background count _____ c/m

Number of sheets	Intensity (c/m)	Corrected Intensity (c/m)
_____	_____	_____
_____	_____	_____
_____	_____	_____
_____	_____	_____
_____	_____	_____
_____	_____	_____
_____	_____	_____
_____	_____	_____
_____	_____	_____

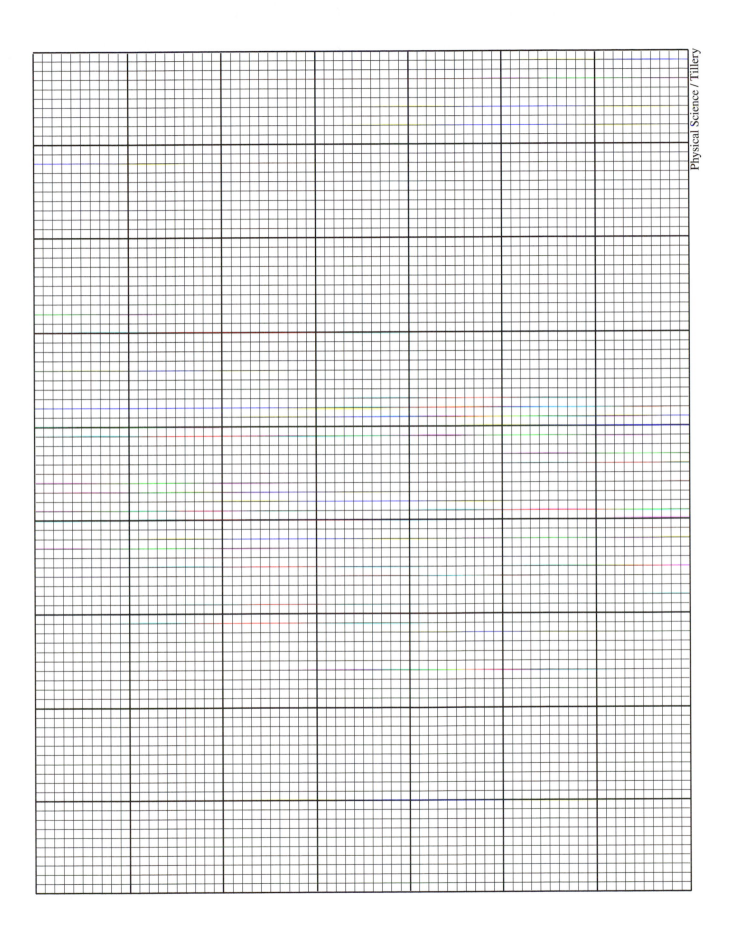

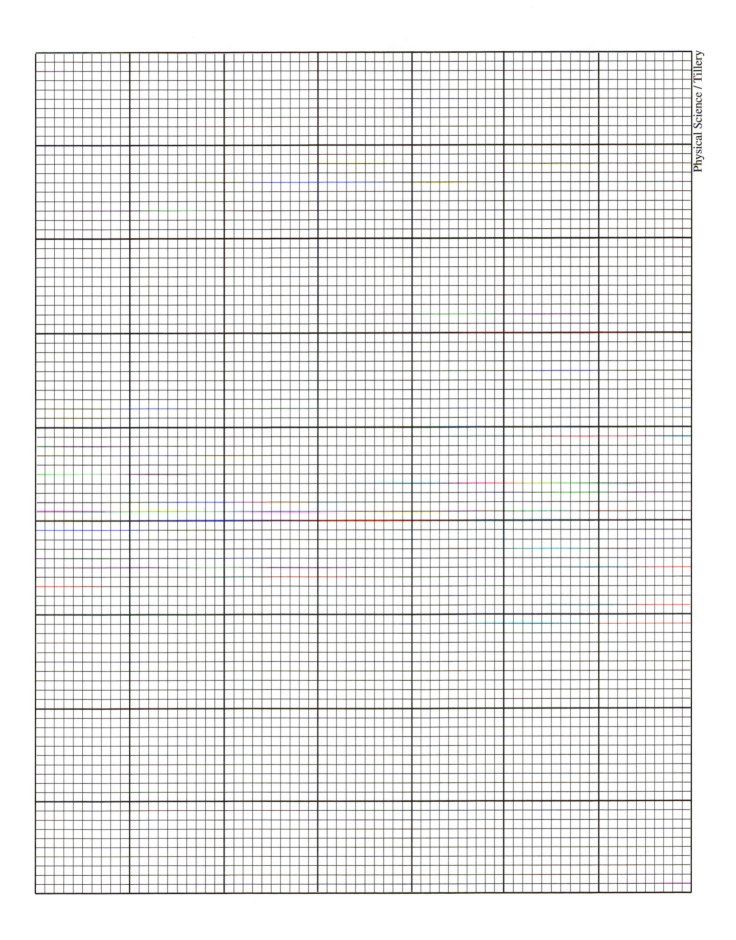

Experiment 41: Growing Crystals

Introduction

A salt dissolves in water because the polar water molecules pull ions away from the crystal lattice of the salt. The oxygen end of a water molecule has a negative polar charge and the hydrogen ends are positive. The oxygen ends of water molecules tend to orient themselves toward the positive ions on the outside of the crystal lattice, as opposite charges attract one another. The hydrogen ends likewise orient themselves toward the negative ions of the lattice. If the attraction of water molecules is greater than the attraction between the ions in the lattice, the ions are pulled away and dissolving occurs. Saturation occurs as water molecules become "tied up" in their attraction for ions. Fewer water molecules means less attraction on the ionic solid, with more solute ions being pulled back to the surface of the crystal lattice.

Each salt forms its own kind of crystal, which is one way to identify salts as well as other crystalline substances such as minerals that occur in the earth's crust. In this investigation a warm, saturated solution is cooled and this lowers the solubility. Rapid cooling—or rapid evaporation of water—results in many small crystals that can be seen with a hand lens. If the conditions are favorable, large beautiful crystals can be "grown" from a saturated solution. This investigation provides a procedure for growing large, beautiful crystals and will require an extended time period.

Materials

Large Pyrex beaker (1 liter or larger), selected salt or salts (see table 41.1 for possibilities and approximate amounts of salt required for each), ring stand and ring, wire screen, burner, balance, glass stirring rod, glass jar (1 liter or larger, with lid), spatula, very fine fishing leader, tape, epoxy glue.

CAUTION: Some salts can be poisonous if taken internally. Wash hands thoroughly after handling salts or solutions.

Procedure

1. Using table 41.1, select a salt to produce a large crystal. Note that several different chemicals may be sold as "alum" by drug and grocery stores. If alum is selected and purchased from a drug or grocery store, make sure it is potassium aluminum sulfate.

2. Using a ring stand and ring, wire screen, and large Pyrex beaker, heat 1 L of distilled water. When the water is hot, begin stirring as you slowly add the approximate amount listed for the salt selected from table 41.1. Stir while slowly adding the salt and continue stirring until no more will dissolve. Increase the solution temperature as necessary, stirring until you are confident that you have a saturated solution. If all the salt dissolves, continue adding small amounts until no more will dissolve.

3. Remove the burner flame, allowing the solution to cool slightly. Decant (pour the solution from any undissolved salts in the bottom) the saturated solution while still warm into a large glass jar. Thread a length of fishing leader through a hole punched in the center of the lid, then tape it securely to the lid. When the lid is on the jar the leader should be long enough to extend halfway into the solution. If the leader floats, use the stirring rod to push it below the surface of the liquid. Place the jar where it can remain undisturbed overnight.

4. Small crystals will form in the solution overnight. Carefully remove the lid and place the fishing leader across a paper towel. Remove all crystals from the leader, selecting the one largest crystal as your seed crystal. Epoxy this seed crystal to the end of the dry leader.

5. Pour the solution from the jar back into the large beaker, being sure that all crystals (except the seed crystal) go with the solution. Heat the solution and again prepare a saturated solution. When you are confident that you have a saturated solution, cool it to room temperature, then return the seed crystal to the solution. Adjust the fishing leader and tape it to the lid so the seed crystal is about centered in the solution. If your seed crystal dissolves, you did not have a saturated solution or the solution was not cooled to room temperature. If you make this error, return to procedure step 3.

6. The solution and seed crystal are now allowed to stand undisturbed, and at a constant temperature, for several days. To produce large, perfect crystals it is absolutely necessary for the solution to be isolated from physical disturbances and at a constant temperature. In addition, any small crystals that form on the surface of the seed crystal during a growth period must be completely removed by breaking, scraping, and sanding.

7. The first growth period is complete when the crystal stops growing. When this occurs remove the crystal from the solution and provide more ions by repeating procedure steps 5 and 6. This process is repeated for more growth periods until the crystal reaches the desired size. Any imperfections that develop are the result of some physical or temperature disturbance.

Results

1. Explain why sudden temperature changes result in irregular growth.

2. Why is it necessary to remove other, smaller crystals from the solution?

3. Why is it necessary to re-saturate the solution for each growth period?

4. Why does each salt form its own kind of crystal?

5. Was the purpose of this lab accomplished? Why or why not? (Your answer to this question should show thoughtful analysis and careful, thorough thinking.)

Invitation to Inquiry

Investigate the crystal structure of some of the large samples. Consider, for example, measuring the angles between the faces, comparing how the crystal transmits light, and other measurements that would help you interpret the structure in terms of atoms or molecules. Does the structure provide any information about the conditions necessary for growing large and well-formed crystals? Or are the same conditions needed for all crystals? Are there any upper limits on the size of a crystal?

Table 41.1	Salts, Crystal Colors, and Amounts Needed to Grow Crystals	
Salt	Crystal Color	Approximate Amount Needed: More than. . .
Potassium aluminum sulfate (alum)	Colorless	225 g
Sodium chloride (table salt)	Colorless	425 g
Copper(II) sulfate (blue vitriol)	Blue	2.02 kg
Nickel sulfate (hydrate)	Blue or Green	3.53 or 4.83 kg
Cobalt(II) sulfate	Red	860 g
Cobalt(II) chloride (hydrate)	Red	1.2 kg
Potassium chromate	Yellow	825 g

Experiment 42: Properties of Common Minerals

Introduction

The earth's crust is made up of rocks, which are solid aggregations of materials that have been cohesively brought together by rock-forming processes. The fundamental building blocks of rocks are **minerals**, which are naturally occurring, inorganic solid elements or chemical compounds with a crystalline structure. About 2,500 different minerals are known to exist, but only about 20 are common in the crust. Examples of these common minerals are quartz, calcite, and gypsum.

Each rock-forming mineral has its own set of physical properties because each mineral has (1) a chemical composition, and (2) a crystal structure that is unlike any other mineral. The exact composition and crystal structure of an unknown mineral can be determined by using laboratory procedures. This type of analysis is necessary when the crystal structures are too small to be visible to the unaided eye. In this laboratory activity you will learn how to identify minerals by considering some identifying characteristics of large, well-developed mineral crystals.

Materials

Mineral collection, streak plate, magnifying glass, steel file, pocketknife, copper penny, magnet, dilute hydrochloric acid, dropper, balance, graduated cylinder and overflow can to determine density (optional).

Procedure

Part A: Developing a Table of Mineral Properties

Examine each mineral specimen in the mineral collection set, one at a time. Consider each of the following properties and record all observations in Data Table 42.1.

1. **Color.** The color of a mineral specimen is an obvious property, but the color of some minerals varies from one specimen to the next because of chemical impurities. Sometimes the property of color can be useful in identifying a mineral. Use combinations of words as necessary to describe mineral colors — for example, reddish-brown.

2. **Streak.** Streak is the color of a mineral when it is finely powdered, a property that is more consistent from specimen to specimen than color. Streak is observed by rubbing the corner of a specimen across an unglazed streak plate. Determine the streak color of each mineral and record your observations in Data Table 42.1. Note that a mineral harder than the streak plate will not leave a streak. This is useful information, so record this observation if it occurs.

3. **Hardness.** Hardness is the resistance of a mineral to being scratched. The Mohs hardness scale is used as a basis for measuring hardness. The hardness test is done by trying to scratch a mineral with a mineral (or some object with an equivalent hardness) from the Mohs scale. If the mineral being investigated scratches a Mohs hardness scale mineral, the mineral in question is harder than the test mineral. If the mineral being investigated is scratched by the test mineral, the mineral in question is not as hard as the test mineral. If both minerals are scratched by each other, they have the same hardness. The Mohs hardness scale is given in table 42.1, along with the hardness of some common objects that can be used for hardness tests. A hardness of 1 on this scale is assigned to the softest mineral, and the hardest mineral has an assigned hardness of 10.

Table 42.1	Mohs Scale of Hardness for Minerals and Common Objects Tests	
Hardness Number	Mineral Example	Common Object Test
1	Talc	Can be scratched with fingernail.
2	Gypsum	Can be scratched with fingernail, but not easily.
3	Calcite	Can be scratched with copper penny.
4	Fluorite	Can be scratched with pocketknife. Will not scratch glass.
5	Apatite	Can be scratched with pocketknife, but not easily.
6	Feldspar	Can be scratched with edge of steel file. Cannot be scratched with pocketknife.
7	Quartz	Scratches glass.
8	Topaz	Scratches quartz.
9	Corundum (ruby or sapphire, for example)	Scratches all minerals but diamond.
10	Diamond	Scratches all minerals. Can be scratched only by another diamond.

4. **Cleavage.** Cleavage is the tendency of a mineral to break along smooth planes. Cleavage occurs in parallel planes in some minerals and in one or more directions in other minerals. Sometimes cleavage is perfect, other minerals may have indistinct cleavage.

5. **Fracture.** Minerals that do not have cleavage may show fracture, an irregular broken surface rather than the smooth planes of cleavage. Some minerals have a distinct way of fracturing, such as the conchoidal fracture of obsidian and quartz. Conchoidal fracture is the breaking along curved surfaces like a shell.

6. **Luster.** Luster describes the surface sheen, the way a mineral reflects light. Minerals that have the surface sheen of a metal are described as being metallic. Nonmetallic luster is described by such terms as pearly (like a pearl), vitreous (like glass), glossy, dull, and so forth.

7. **Density.** Density is a ratio of the mass of a mineral to its volume. Often the density of a mineral is expressed as specific gravity, which is a ratio of the mineral density to the density of water. To obtain an exact specific gravity, a mineral specimen must be pure and without cracks, bubbles, or substitutions of chemically similar elements.

8. **Other properties.** There are a few other properties that are specific for one or more minerals. These properties are determined by tests such as a reaction to acid, reaction to a magnet, and others. Some of the specific properties of certain minerals, such as the double image seen through a calcite crystal, are unique enough to identify an unknown mineral in an instant.

Part B: Unknown Minerals

Once the Table of Mineral Properties (Data Table 42.1) is complete, you can use it to identify an unknown mineral. The procedure is find out what the mineral is not. For example, suppose you start with the streak test and find the unknown mineral leaves a red streak. This test tells you the unknown mineral is not one of the minerals that leave some other color of streak. Then you look at the minerals that have a red streak and perform a second test. The second test eliminates still other possibilities. Eventually, you find what the unknown mineral is by finding out what it isn't.

Results: Was the purpose of this lab accomplished? Why or why not? (Your answer to this question should show thoughtful analysis and careful, thorough thinking.)

Data Table 42.1	Table of Mineral Properties		
Mineral	Color	Luster	Streak
Calcite			
Magnetite			
Pyrite			
Hematite			
Limonite			
Quartz			
Fluorite			
Apatite			
Feldspar			
Talc			
Hornblende			
Halite			
Bauxite			
Biotite			
Muscovite			
Sphalerite			
Galena			
Chrysotile			
Gypsum			

Mineral	Hardness	Cleavage/Fracture	Other
Calcite			
Magnetite			
Pyrite			
Hematite			
Limonite			
Quartz			
Fluorite			
Apatite			
Feldspar			
Talc			
Hornblende			
Halite			
Bauxite			
Biotite			
Muscovite			
Sphalerite			
Galena			
Chrysotile			
Gypsum			

Data Table 42.1   Table of Mineral Properties, Continued

Experiment 43: Density of Igneous Rocks

Introduction

Igneous rocks are rocks that form from the cooling of a hot, molten mass of rock material. Igneous rocks, as other rocks, are made up of various combinations of minerals. Each mineral has its own temperature range at which it begins to crystallize, forming a solid material. Minerals that are rich in iron and magnesium tend to crystallize at high temperatures. Minerals that are rich in silicon and poor in iron and magnesium tend to crystallize at lower temperatures. Thus minerals rich in iron and magnesium crystallize first in a deep molten mass of rock material, sinking to the bottom. The minerals that crystallize later will become progressively richer in silicon as more and more iron and magnesium are removed from the melt.

Igneous rocks that are rich in silicon and poor in iron and magnesium are comparatively light in density and color. The most common igneous rock of this type is **granite**, which makes up most of the earth's continents. Igneous rocks that are rich in iron and magnesium are dark in color and have a relatively high density. The most common example of these dark-colored, more dense rocks is **basalt**, which makes up the ocean basins and much of the earth's interior. Basalt is also found on the earth's surface as a result of volcanic activity.

Materials

Balance, overflow can, graduated cylinder, ring stand and ring, thin nylon string, beaker, granite specimen, and basalt specimen.

Procedure

1. Use a balance to find the mass of a basalt rock. Record the mass in Data Table 43.1. Tie a 20-cm length of nylon string around the rock so you can lift it with the string. Test your tying abilities to make sure you can lift the rock by lifting the string without the rock falling.

2. Place an overflow can on a ring stand, adjusted so the overflow spout is directly over a graduated cylinder.

3. Hold a finger over the overflow spout, then fill the can with water. Remove your finger from the spout, allowing the excess water to flow into the cylinder. Dump this water from the cylinder, then place it back under the overflow spout.

4. Grasp the free end of the string tied around the basalt, then lower the rock completely beneath the water surface in the overflow can. The volume of water that flows into the graduated cylinder is the volume of the rock. Remembering that a volume of 1.0 mL is equivalent to a volume of 1.0 cm^3, record the volume of the rock in cm^3 in Data Table 43.1.

5. Calculate the mass density of the basalt and record the value in the data table.

6. Repeat procedure steps 1 through 5 with a sample of granite.

Results

1. In what ways do igneous rocks have different properties?

2. Explain the theoretical process or processes responsible for producing the different properties of igneous rocks.

3. According to the experimental evidence of this investigation, propose an explanation for the observation that the bulk of the earth's continents are granite, and that basalt is mostly found in the earth's interior.

4. Was the purpose of this lab accomplished? Why or why not? (Your answer to this question should show thoughtful analysis and careful, thorough thinking.)

Invitation to Inquiry

Survey the use of rocks used in building construction in your community. Compare the type of rocks that are used for building interiors and those that are used for building exteriors. Are any trends apparent for buildings constructed in the past and those built in more recent times? If so, are there reasons (cost, shipping, other limitations) underlying a trend or is it simply a matter of style?

Data Table 43.1	Density of Igneous Rocks		
	Mass (g)	Volume (cm^3)	Density (g/cm^3)
Basalt	_____ g	_____ cm^3	_____ g/cm^3
Granite	_____ g	_____ cm^3	_____ g/cm^3

Experiment 44: Latitude and Longitude

Introduction

The continuous rotation and revolution of the earth establish an objective way to determine directions and locations on the earth. If the earth were an unmoving sphere there would be no side, end, or point to provide a referent for directions and locations. The earth's rotation, however, defines an axis of rotation which serves as a reference point for determination of directions and locations on the entire surface. The reference point for a sphere is not as simple as on a flat, two-dimensional surface, because a sphere does not have a top or side edge. The earth's axis provides the north-south reference point. The equator is a big circle around the earth that is exactly halfway between the two ends, or poles of the rotational axis. An infinite number of circles are imagined to run around the earth parallel to the equator. The east- and west-running parallel circles are called **parallels**. Each parallel is the same distance between the equator and one of the poles all the way around the earth. The distance from the equator to a point on a parallel is called the **latitude** of that point. Latitude tells you how far north or south a point is from the equator by telling you on which parallel the point is located.

Since a parallel is a circle, a location of 40° N latitude could be anyplace on that circle around the earth. To identify a location you need another line, one that runs pole to pole and perpendicular to the parallels. North-south running arcs that intersect at both poles are called **meridians**. There is no naturally occurring, identifiable meridian that can be used as a point of reference such as the equator serves for parallels, so one is identified as the referent by international agreement. The reference meridian is the one that passes through the Greenwich Observatory near London, England, and is called the **prime meridian**. The distance from the prime meridian east or west is called the **longitude**. The degrees of longitude of a point on a parallel are measured to the east or to the west from the prime meridian up to 180°.

Locations identified with degrees of latitude north or south of the equator and degrees of longitude east or west of the prime meridian are more precisely identified by dividing each degree of latitude into subdivisions of 60 minutes (60') per degree, and each minute into 60 seconds (60"). In this investigation you will do a hands-on activity that will help you understand how latitude and longitude are used to locate places on the earth's surface.

Materials

Soft clay (fist-sized lump), protractor, toothpicks, knife, pencil.

Procedure

1. Obtain a lump of clay about the size of your fist. Knead the clay until it is soft and pliable, then form it into a smooth ball for a model of the earth.

2. Obtain a sharpened pencil. Hold the clay ball in one hand and use a twisting motion to force the pencil all the way through the ball of clay. Reform the clay into a smooth ball as necessary. This pencil represents the earth's axis, an imaginary line about which the earth rotates. Hold the clay ball so the eraser end of the pencil is at the top. The eraser end of the pencil represents the North Pole and the sharpened end represents the South Pole. With the North Pole at the top, the earth turns so the part facing you moves from left to right. Hold the clay ball with the pencil end at the top and turn the ball like this to visualize the turning earth.

3. The earth's axis provides a north-south reference point. The equator is a circle around the earth that is exactly halfway between the two poles. Use the end of a toothpick to make a line in the clay representing the equator.

4. Hold the clay in one hand with the pencil between two fingers. Carefully remove the pencil from the clay with a back and forth twisting motion. Reform the clay into a smooth ball if necessary, being careful not to destroy the equator line. Use a knife to slowly and carefully cut halfway through the equator. Make a second cut down through the North Pole to cut away one-fourth of the ball as shown in figure 44.1. Set the cut-away section aside for now.

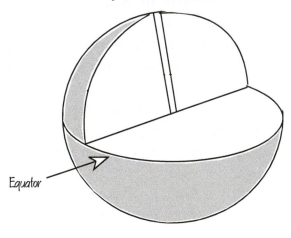

Figure 44.1

5. Place a protractor on the clay ball where the section was removed. As shown in figure 44.2, the 0° of the protractor should be on the equator and the 90° line should be on the axis (the center of where the pencil was). You may have to force the protractor slightly into the clay so the 0° line is on the equator. Directly behind the protractor, stick toothpicks into the surface of the clay ball at 20°, 40°, and 60° above the equator on both sides. Remove the protractor from the clay and return the cut-away section to make a whole ball again.

6. Use the end of a toothpick to make parallels at 20°, 40°, and 60° north of the equator, then remove the six toothpicks. Recall that parallels are east and west running circles that are the same distance from the equator all the way around the earth (thus the name parallels). The distance from the equator to a point on a parallel is called the **latitude** of that point. Latitude tells you how far north or south a point is from the equator. There can be 90 parallels between the equator and the North Pole, so a latitude can range from 0° North (on the equator) up to 90° North (at the North Pole).

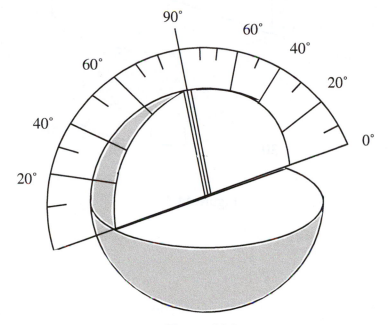

Figure 44.2

7. Since a parallel is a circle that runs all the way around the earth, a second line is needed to identify a specific location. This second line runs from pole to pole and is called a meridian. To see how meridians identify specific locations, again remove the cut section from the ball of clay. This time place the protractor flat on the equator as shown in figure 44.3, with the 90° line perpendicular to the axis. Stick toothpicks directly below the protractor at 0°, 60°, 120°, and 180°, then remove the protractor and return the cut-away section to make a whole ball again. Use a toothpick to draw lines in the clay that run from one pole, through the toothpicks, then through the other pole. By agreement, the 0° line runs through Greenwich near London, England and this meridian is called the prime meridian. The distance east or west of the prime meridian is called **longitude**. If you move right from the prime meridian you are moving east from 0° all the way to 180° East. If you move left from the prime meridian you are moving west from 0° all the way to 180° West.

8. Use a map or a globe to locate some city that is on or near a whole number latitude and longitude. New Orleans, Louisiana, for example, has a latitude of about 30° N of the equator. It has a longitude of about 90° W of the prime meridian. The location of New Orleans on the earth is therefore described as 30° N, 90° W. Locate this position on your clay model of the earth and insert a toothpick. Compare your model to those of your classmates.

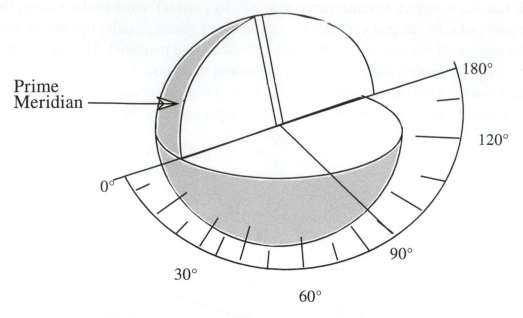

Figure 44.3

Results

1. What information does the latitude of a location tell you?

2. What information does the longitude of a location tell you?

3. According to a map or a globe, what is the approximate latitude and longitude of the place where you live?

4. Explain how minutes and seconds are used to identify a location more precisely.

5. Was the purpose of this lab accomplished? Why or why not? (Your answer to this question should show thoughtful analysis and careful, thorough thinking.)

Invitation to Inquiry

There are several ways to find your latitude by measurement. First, determine your latitude by measuring the angle of the North Star above the horizon. Second, determine your latitude by measuring the angle between a vertical stick and a line to the noonday sun on the spring equinox (March 21) or the autumnal equinox (September 23). For the North Star, consider making two measurements 12 hours apart and averaging the two. Why do these two different methods tell you your latitude? Is one more in "agreement" with the stated latitude for your location?

Experiment 45: Topographic Maps

Introduction

A map is a convenient representation of an area that describes basic features and locations. A **road map** show roads, political boundaries, cities, and natural features such as rivers and lakes. A **topographic map** shows the topography—the three-dimensional shape of the surface, including mountains, hills, and valleys—in addition to roads, buildings, dams, and political boundaries. **Geologic maps** show the distribution of different rock bodies exposed at the earth's surface, and frequently include features found on topographic maps.

Most maps are located with respect to the reset of the earth by a latitude-longitude coordinate system of lines. This system uses east-west lines called lines of latitude, or parallels, and north-south lines called lines of longitude, or meridians (see Experiment 44).

A topographic map shows elevation difference by means of *contour lines*. Figure 45.1 shows how the topography of the area sketched in the top diagram is depicted with contour lines in the bottom diagram. Contour lines on the bottom diagram are draw at intervals of 20 feet, starting with 0 at mean sea level.

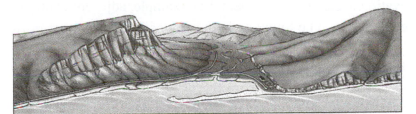

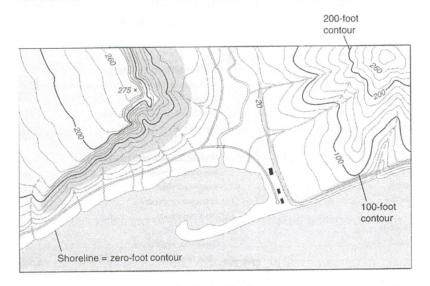

Figure 45.1

Source: U.S. Geological Survey

The following characteristics of contour lines, most of which are illustrated in figure 45.1, govern the construction and reading of contour maps:

1. Every point on the same contour line has the same elevation.
2. A contour line always rejoins itself to form a closed loop, although this may or may not occur within the map area. Thus, if you walked along a contour, you would eventually get back to your starting point.
3. Contour lines never split.
4. Contour lines never cross one another.
5. Slopes rise of desend at right angles to any contour line.
   - Evenly spaced contours indicate a uniform slope
   - Closely spaced contours indicate a steep slope
   - Widely spaced contours indicate a gentle slope
   - Unevenly spaced contours indicate an irregular slope
6. Contours usually encircle a hilltop; if the hill falls within the map area, the highest point will be inside the innermost contour.
7. Contour lines near the tops of ridges or bottoms of valleys always occur in pairs having the same elevation on either side of the ridge or valley.
8. Contours always bend upstream when they cross valleys.
9. If two or more adjacent contour lines have the same elevations, a change of elevation occurs between them. For example, adjacent contours with the same elevation would be found on both sides of a valley bottom or ridge top.

Materials

Soft clay (two fist-sized lumps), clear plastic shoe box (or other clear plastic box), toothpick or sharp pencil, metric ruler, knife, masking tape, marking pen, sheet of clear acetate (such as overhead projector film).

Procedure

1. Use two fist-sized lumps of modeling clay to build a small model of a mountain inside one end of a clear plastic shoe box. The top of the mountain should be about 6-cm taller than the lowest point.

2. Carve a V-shaped valley from the base of the mountain to the other end of the box.

3. Place a piece of masking tape vertically, from top to bottom anywhere on the outside of the shoe box. Measure and mark 1-cm intervals, beginning with 0 near the bottom and the 8-cm end near the top.

4. Add enough water to fill the container to the 0 mark on the tape. This mark represents sea level.

5. Use a toothpick or sharp pencil to trace the water level on your model.

6. Add more water to the shoe box until the water level reaches the 1-cm mark. Again trace the water level on the model.

7. Continue to increase the water level by 1-cm increments, tracing the water level each time. Continue until the water level covers the model or is near the top of the shoe box. Then empty the water from the box.

8. Place the lid on the shoebox. Put a sheet of overhead projector acetate on the lid. Look straight down on the top of the model and trace the contour lines with a marking pen. You will make a contour line map of your model.

9. Without showing the models, trade maps with a classmate. Use the list of contour line characteristics from the introduction to this lab to interpret your classmate's contour map.

Results

1. What do topographic maps show?

2. What is the relationship between the closeness of contour lines and the steepness of slopes?

347

3. What happens to contour lines across a valley?

4. How would a nearly vertical cliff be represented on a contour map?

5. Was the purpose of this lab accomplished? Why or why not? (Your answer to this question should show thoughtful analysis and careful, thorough thinking.)

Go to www.lib.utexas.edu/maps/national-parks.html. Select Devils Tower National Monument [Wyoming] shaded relief map and answer the following:

(1.) What is the elevation of Devils Tower?

(2.) What is the contour interval of the map?

(3.) What is the approximate height of Devils Tower?

(4.) What is the top of Devils Tower like?

(5.) Is Devils Tower too steep to hike to the top?

Experiment 46: Telescopes

Introduction

A convex lens can be used as a "burning glass" by moving the lens back and forth until the sunlight is focused into a small bright spot—an image of the sun. This image is hot enough to scorch the paper, perhaps setting it on fire. The lens is moved back and forth to refract the parallel rays of light from the sun the point where the image is formed on the paper. The place where the image forms is called the **focal point** of the lens. The distance from the focal point to the lens is called the **focal length** (f). The focal length of a lens is determined by its index of refraction and the shape of the lens. The focal length is an indication of the refracting ability, or strength of a lens. A lens with a short focal length is considered to be a stronger lens than one with a longer focal length.

There are three important measurements that are used to describe how lenses work as optical devices. These are (1) the *focal length* (f), (2) the *image distance* (d_i), the distance from the lens that an image is formed, and (3) the *object distance* (d_o), the distance from the object being imaged to the lens. The relationship between these measurements is given in the **lens equation**, which is

$$\frac{1}{f} = \frac{1}{d_o} + \frac{1}{d_i}$$

The magnification produced by a lens is defined as the ratio of the height of the image (h_i) to the height of the object (h_o). This is also equal to the ratio of the image distance (d_i) to the object distance (d_o), or

$$\text{Magnification} = \frac{d_i}{d_o}$$

In this investigation you will compare measuring the focal length of a lens directly with using the lens equation to calculate the focal length of a convex lens. Magnification of a lens will also be investigated by comparing direct measurement of magnification with theoretical magnification as calculated from the focal lengths of two lenses.

Procedure

1. Measure the focal length of a lens directly. First, place a lens in a lens holder and secure it at the 50 cm mark on a meter stick. Second, point the meter stick at some distant objects, such as a tree or a house about a block away. Move a cardboard screen in a holder back and forth until you obtain a sharp image on the screen. Finally, measure the distance between the sharp image and the lens, which is the focal length (f) of the lens. Record the focal length in Data Table 46.1 on page 350.

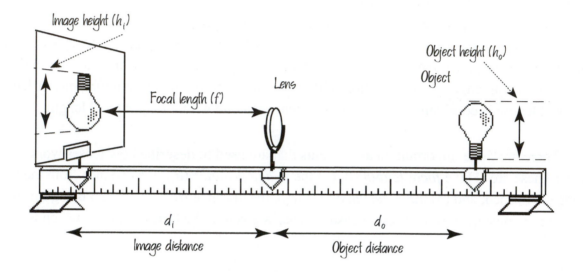

Figure 46.1

2. Find the focal length of the lens used in procedure step 1 by use of the lens equation. Set up a meter stick in holders, a screen in a holder, and a luminous object in a holder as shown in figure 46.1. The room should be darkened, then place the screen at the focal length distance found in procedure step 1. With the screen and object fixed in place, slowly move the lens along the meter stick to obtain the sharpest image possible. Note that the object and image should lie on a straight line along, and perpendicular to the principal axis of the lens. Measure and record in Data Table 46.1 the object distance (d_o) and the image distance (d_i) to the nearest 1 mm. Measure and record to the nearest 0.5 mm the height of the object (h_o) and height of the image (h_i). Record other observations here.

3. Place the lens *less than one focal length* from the bulb, then move the screen back and forth to see if you can obtain an image on the screen. Look through the lens at the bulb. Record your observations here:

4. Repeat procedure steps 1 and 2 for three more lenses. Record all data and results of calculations in Data Table 46.1. Select one of the lenses with a short focal length (for eyepiece lens) and one with a longer focal length (for objective lens) for use in the next procedure step.

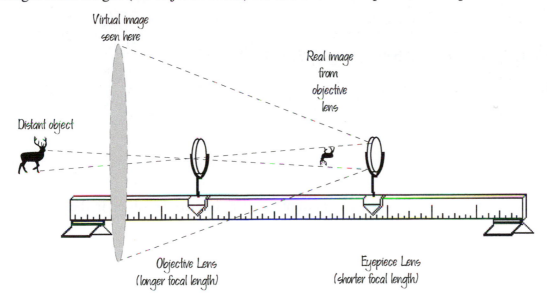

Figure 46.2

5. Make a telescope by mounting a short and longer focal length lenses on a meter stick as shown in figure 46.2. Focus the telescope by adjusting the shorter focal length lens until it magnifies the image from the longer focal length lens. Calculate the *theoretical* magnification of your telescope by dividing the focal length of the objective lens by the focal length of the eyepiece lens. Record the theoretical magnification in Data Table 46.2.

6. Work with a partner to *determine experimentally* the magnification of your telescope. Focus your telescope on some object in front of your partner, who should hold a meter stick next to the object. Look at the object normally with one eye and look at the object through the telescope with the other eye. Direct your partner to position a pointer on the meter stick to indicate the apparent size of the enlarged image. Record the height of the image and the height of the object in Data Table 46.2, then calculate the magnification.

Results

1. What relationships did you find between d_o, d_i, and f?

2. What relationships did you find between d_o, d_i, h_i and h_o?

3. Discuss the advantages, disadvantages, and possible sources of error involved in the two ways of finding the focal length of lenses.

4. Discuss the advantages, disadvantages, and possible sources of error involved in the two ways of finding the magnification of lenses.

5. Was the purpose of this lab accomplished? Why or why not? (Your answer to this question should show thoughtful analysis and careful, thorough thinking.)

Invitation to Inquiry

Design an experiment to study the effect of the diameter of a lens on the image formed. Do the experiment.

Use plane (flat), concave, and convex mirrors to find when you can see:

- an enlarged image.
- a reduced image.
- an image of the same size.
- an image that appears upright.
- an image that appears inverted.
- an image that appears upright, then inverted after some adjustment.

What generalizations can you make to inform someone how to make the various images with mirrors?

Lens	Focal length measured directly (f) (cm)	Object distance (d_o) (cm)	Image distance (d_i) (cm)	Focal length from lens equation (f) (cm)	Object size (h_o) (cm)	Image size (h_i) (cm)	Magnification
1	_____	_____	_____	_____	_____	_____	_____
2	_____	_____	_____	_____	_____	_____	_____
3	_____	_____	_____	_____	_____	_____	_____
4	_____	_____	_____	_____	_____	_____	_____

Data Table 46.1   Lens Focal Length and Magnification

Data Table 46.2	Lens Magnification	
Objective lens focal length		_____
Eyepiece lens focal length		_____
Theoretical magnification (Focal length of objective ÷ focal length of eyepiece)		_____
Object height (h_o)		_____
Image height (h_i)		_____
Experimental magnification ($h_i \div h_o$)		_____

Experiment 47: Celestial Coordinates

Introduction

To an observer unencumbered by scientific models, the night sky appears to be an inverted bowl resting on a flat plane. The observer appears to be located at the center of the bowl. Sprinkled over the inner surface of the bowl are the stars arranged in fixed, identifiable patterns that do not change noticeably from day to day, year to year, or even century to century.

Among what ancient observers called the "fixed stars" seven objects move: the sun, the moon, and five apparently star-like objects called planets. Other objects that moved in the sky but were transitory, such as meteors and comets, were considered by the ancient observers, Aristotle in particular, to be atmospheric phenomena and were not considered in modeling the heavens.

The motion of the sun was readily apparent. The general direction of its rising is synonymous with what we call the east. The direction of its setting is the west. Each day the sun rises to its highest altitude above the horizon when it is due south. If we imagine a line extending from the north point on the horizon passing overhead and continuing to the south point on the horizon, then the sun reaches its highest altitude when it crosses this line called the **meridian**. When the sun, or any other object in the sky, crosses the meridian, the object is said to **transit.**

During the course of a single night, the stars wheel around a fixed point as if the celestial bowl on which they are seemingly attached were spinning on an axis passing through this point and the observer's position. On a time exposure photograph made with a camera pointed toward this fixed point, each star traces out an arc of a perfect circle. Careful observation shows that the bowl rotates from east to west at a uniform rate, completing one revolution in just under 24 hours (23h56m4.s091). That is, the time interval between successive meridian transits of a particular star, say, is a bit shorter than the average time between successive meridian transits of the sun, 24 hours exactly. (The word "average" is used because the time interval between successive solar transits varies throughout the year.)

Since the sun rate and the star rate are close but not quite the same, it must be that the sun moves slowly with respect to the background stars, and in fact, the sun's path among the stars can be delineated. By noticing the sun's position among the stars at sunrise or sunset each day, we can plot its path among the stars. This path turns out to be a circle and is called the **ecliptic**, and the sun creeps from *west to east* at a nearly uniform rate, taking 365 days and almost 6 hours to complete a single circuit. Where is this circle on the celestial sphere? In order to answer this question, we will establish a coordinate system *on the sky* so that we can specify locations of points and circles. We will produce an analog of the system of latitude and longitude used to locate positions on the earth's surface.

Procedure

Part A: The Celestial Pole and Equator

Let us call the fixed point about which the sphere of stars turns the celestial pole. The direction on the ground toward this fixed point is called *north* if we are in the northern hemisphere of the earth. If we are in the southern hemisphere, a different fixed point is apparent, and the direction toward the point is *south*. In the northern sky a bright star, Polaris, is near the north celestial pole. There is no bright star near the south celestial pole.

On the surface of the earth, the line that is everywhere 90° away from the pole is called the equator. It is a circle bisecting the earth halfway between each pole. In the sky, the line (circle) on the celestial sphere that is everywhere 90° from the celestial pole is called the celestial equator. Figure 47.1 shows the geometry of the celestial sphere relative to that of the earth. The angle α formed by the observer, the center of the earth, and a point on the equator is defined to be the observer's latitude on the earth.

1. Prove that the angle α, the observer's latitude, is in fact the same as the angle (alpha) of the north celestial pole over the northern horizon.

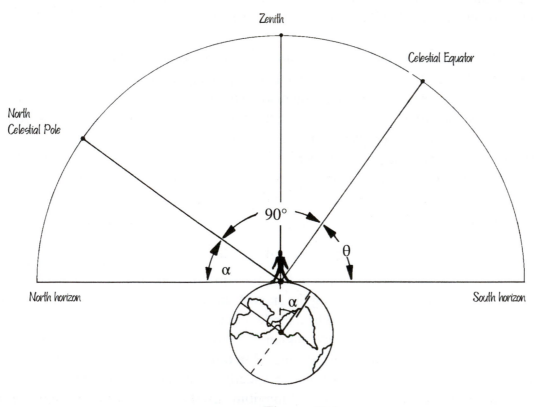

Figure 47.1

If the celestial pole is an angle α above the northern horizon, then we can establish the location of the celestial equator. From fig. 47.1, the angle along the meridian that the equator makes over the southern horizon is θ = 180° − 90° − α. So the celestial equator starts in the east, climbs up to an angle of 90°− α over the southern horizon, and then drops down to the west point of the horizon.

2. If the celestial pole makes an angle of 40° over the northern horizon, what angle over the southern horizon does the equator make?

3. Suppose you are at Quito, Ecuador (latitude = 0°). Where is the celestial equator?

The ecliptic is a circle on the celestial sphere that is tipped at an angle of 23 1/2° to the celestial equator. Figure 47.2 illustrates the relationship between the celestial equator and the ecliptic. Since during the course of a year the sun traces out the path of the ecliptic, it appears at different places with respect to the celestial equator. At one time, the sun is maximally above (north of) the celestial equator, sometimes right on the equator, and at another time maximally below (south of) the equator. Figure 47.3 shows two different orientations of the sun with respect to the celestial equator. The figures are drawn for the case where the north celestial pole lies at an angle of 40° over the north point of the horizon (i.e., latitude = 40° north). The figures are also drawn for noon, when the sun is on the meridian.

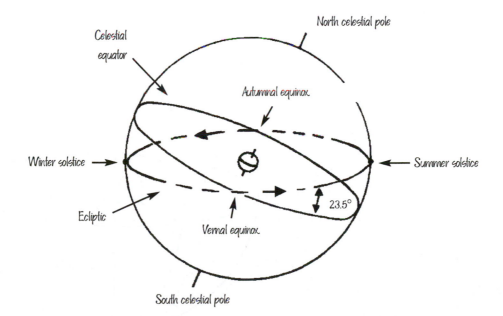

Figure 47.2 The ecliptic (dashed line) intersects the celestial equator at two points call equinoxes. The solstices mark the most northerly and southerly points.

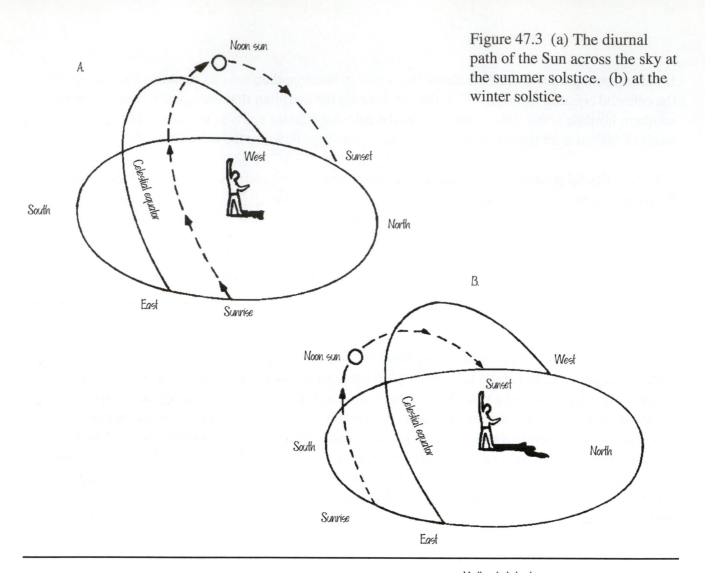

Figure 47.3 (a) The diurnal path of the Sun across the sky at the summer solstice. (b) at the winter solstice.

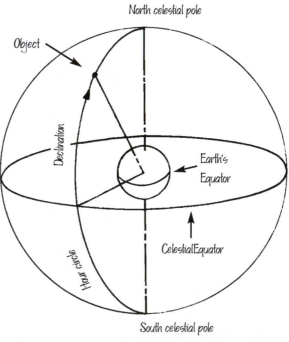

Figure 47.4 The declination of an object on the celestial sphere is its angular distance measured in degrees north (+) or south (–) of the celestial equator.

Figure 47.5 The right ascension of an object on the celestial sphere.

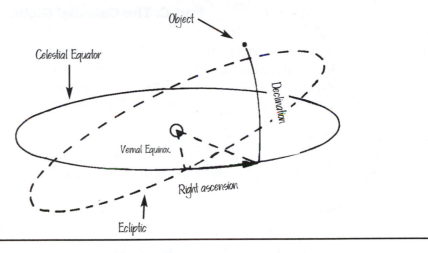

Object

Celestial Equator

Declination

Vernal Equinox

Right ascension

Ecliptic

Part B: Celestial Coordinates

We can now use the points and circles we have discussed thus far to establish a coordinate system to map the heavens. In analogy with the latitude-longitude system used to specify the location of a point on the earth's surface, we will need two coordinates, called **declination** and **right ascension** to locate a particular point in the sky. The declination of a star is its angular distance measured in degrees between the celestial equator and the point or celestial object. The object's declination is positive (+) if it is north of the celestial equator and negative (−) if it is south. Subdivisions are measured in the usual minutes and seconds of arc. Figure 47.4 illustrates this coordinate. Points on the celestial equator have a declination of 0°; the north celestial pole as dec = +90°; the south pole has dec = −90°.

The right ascension coordinate is a bit more complicated. The right ascension of an object is the angular distance measured along the celestial equator between the vernal equinox and the point on the equator that intersects an hour circle passing through the object (see fig. 47.5). An hour circle is a great circle that passes through both celestial poles as well as through the object itself.

Here is the peculiar aspect of the right ascension coordinate: it is measured in **hours**, with subdivisions of minutes and seconds of **time**. Instead of 360° all the way around the celestial equator, right ascension has 24h, that is 24h is equivalent to 360°. Right ascension is measured eastward from the vernal equinox.

4. How many degrees of arc are equivalent to 1^h of right ascension?

5. How many minutes of right ascension are equivalent to 1° of arc?

6. How many minutes of arc are equivalent to 1^m of right ascension?

Part C: The Celestial Globe

A celestial globe is a representation of the celestial sphere, the stars or other objects in the heavens, together with the coordinate system we have been discussing. Remember, the celestial sphere has you, the observer, at its center, so everything in the celestial globe is plotted from this perspective. As you look at the globe, you will have to visualize that you are at the center of the globe looking out.

7. Locate on the globe the (1) north and south celestial poles, (2) the celestial equator, (3) the ecliptic, (4) the two equinoxes where the ecliptic and equator intersect, and (5) the solstices.

8. Which stars are located at the following coordinates?

Right Ascension	Declination	Star Name
$6^h 43^m$	$-16°\ 39'$	
$6^h 22^m$	$-52° 40'$	
$18^h 35^m$	$+38°\ 44'$	
$5^h 12^m$	$-8° 15'$	
$7^h 37^m$	$+5°\ 21'$	
$14^h 36^m$	$-60° 38'$	

9. Which constellation is located at the approximate positions indicted?

Right Ascension	Declination	Constellation
11^h	$+50°$	
19^h	$-25°$	
3^h	$+20°$	
1^h	$+60°$	
13^h	$-50°$	
7^h	$-40°$	

The celestial globe is pivoted about the celestial poles and is held in place by a vertical circular ring of metal that is graduated in degrees. A large horizontal ring serves as a representation of the horizon. Adjust the vertical ring so that the celestial pole is at the proper altitude over the horizon for your latitude. (You may need to ask the laboratory instructor for your latitude to the nearest degree.)

As you rotate the globe you will notice that the celestial equator always maintains a single orientation in the sky; it comes up out of the east, reaches a maximum altitude when it crosses the meridian (vertical metal circle), and sinks in the west.

Part D: Relative Amounts of Daylight and Darkness

On the ecliptic locate the position of the sun on June 21. Place a small piece of 3M self-stick paper to mark the location. *Do not write on the globe or make any permanent markings.* Note the sun's declination for this day and record it in table 47.1. Now rotate the globe until the sun is in its rising position on the eastern horizon.

10. Does the sun rise exactly in the east on this day?

Rotate the globe until the sun sets and note where on the horizon the sun disappears below the horizon. Bring the sun back to the rising position and by counting how many right ascension hour circles lie between the rising and setting points for the sun, estimate to the nearest 15 minutes the amount of time the sun is above the horizon on June 21. Enter your estimate in table 1. By subtracting this number from 24^h, calculate the length of darkness for this date.

Repeat this for December 21 and March 21.

Readjust the position of the celestial pole to correspond to an observer at Altengaard, Norway (latitude = 70° north). Fill in table 47.2 as you did table 47.1.

Readjust the globe one more time to correspond to an observer in Quito, Ecuador (latitude = 0°). Fill in table 47.3.

11. *Calculate* the altitude of the sun (in degrees) over the southern horizon at noon on June 21 for an observer at latitude +40° N. Check your answer using the celestial globe.

12. Repeat procedure 11 for December 21.

Table 47.1 Relative Amounts of Daylight and Darkness for Your Latitude at Different Times of the Year

Date	Sec. of Sun	Number of hours of Daylight	Number of Hours of Darkness
June 21			
December 21			
March 21			

Table 47.2 Relative Amounts of Daylight and Darkness for Altengaard, Norway (Latitude = 70° N) at Different Times of the Year

Date	Sec. of Sun	Number of hours of Daylight	Number of Hours of Darkness
June 21			
December 21			
March 21			

Table 47.3 Relative Amounts of Daylight and Darkness for Quito, Ecuador (Latitude = 0°) at Different Times of the Year

Date	Sec. of Sun	Number of hours of Daylight	Number of Hours of Darkness
June 21			
December 21			
March 21			

13. Eratosthenes, who lived in the second century B.C., was a Renaissance man before his time. He was an astronomer, a geographer, a historian, a mathematician, and a poet. With such a diverse background, it is not surprising that he was the director of the Great Library of Alexandria. In one of the scrolls at the library, Eratosthenes read that at noon on June 21 in the southern frontier outpost of Syene, Egypt, near the first cataract of the Nile, obelisks cast no shadows. As noon approached, the shadows of temple columns grew shorter until at noon they were gone. Reportedly, a reflection of the sun could then be seen at the bottom of a well. On June 21 at Syene, the sun was directly overhead. Use the celestial globe to determine the latitude of Syene.

14. Readjust the celestial globe for your latitude. What range of declinations can stars have so that they never set, that is, are always above the horizon?

Your instructor will provide you with the coordinates of Mercury and Venus for the day of this lab session. Put marked pieces of 3M self-stick paper at the corresponding locations on the celestial globe. Locate the sun for today and place a piece of paper at the sun's location on the globe.

15. When is Mercury visible in a dark sky? (circle one)

> just after sunset
>
> just before sunrise

16. When is Venus visible in a dark sky?

> just after sunset
>
> just before sunrise

17. Estimate how long each is visible in a dark sky.

> Mercury _____
>
> Venus_____

Experiment 48: Motions of the Sun

Introduction

(Note: Experiment 47, "Celestial Coordinates"on page 359 is a prerequisite for this exercise). The sun rises above the eastern horizon, traces an arched path across the sky, and then sinks below the western horizon. Midway between sunrise and sunset the sun climbs to its highest altitude over the horizon in the south. This daily event, the transit of the sun across the celestial meridian, defines **noon**, a fundamental reference in the measurement of time. The interval from one noon to the next sets the length of the **solar day**. Subdivisions of the day have had a long and sordid history, and not everyone welcomed the partitioning of the day into smaller units. As Platus lamented c. 200 B.C.:

> The gods confound the man who first found out how to distinguish hours! Confound him too, who in this place set up a sundial to cut and hack my days so wretchedly into small pieces.

The Romans were the principal partitioners of the day. By the end of the fourth century B.C., they formally divided their day into two parts: before midday (*ante meridiem*, A.M., L. before the meridian) and after midday (*post meridiem*, P.M., L. after the meridian). An assistant to the Roman consul was assigned the task of noticing when the sun crossed the meridian and announcing it in the Forum, since lawyers had to appear in the courts before noon.

By the beginning of the Common Era, the Romans eventually made finer subdivisions of the day. The "hours" of their daily lives were one-twelfth of the time of daylight or of darkness. These variant "hours"—equal subdivisions of the total time of daylight or of darkness—were quite elastic and not really chronometric hours at all. For example, at the time of the winter solstice, by our modern measures there would be only 8 hours, 54 minutes of daylight, leaving 15 hours, 6 minutes for darkness. Near the winter solstice, the Roman daylight "hour" corresponds to

$$\frac{1}{12}(8^{\mathrm{h}}54^{\mathrm{m}}) = 45\frac{1}{2} \text{ minutes}$$

by modern measure.

**Calculate the length (in modern time units) of a Roman "hour" at night at the time of the winter solstice.

At the summer solstice the times were exactly reversed. One-twelfth of a changing interval of time was not a constant from day to day. These "hours" came to be called "temporary hours" or "temporal hours," for they had meaning and length that was only temporary and did not equal an hour the next day. From the Romans' point of view, both day and night always had precisely 12 hours year round. What a problem for the clockmaker!

Sundials were common and were a universal measure of time. They were handy measuring devices since a simple sundial could be made anywhere by anybody without much in the way of special knowledge or equipment. But the cheery boast "I count only the sunny hours" inscribed on many modern sundials, also announces the obvious limitation of the sundial for measuring time. A sundial measures the position of the sun's shadow; thus, no sun, no shadow. And what do you use at night? By the Roman era, water clocks became prevalent and served as a way to measure the shadowless and dark hours. Such clocks had a limited precision, at least by modern standards, but we must be amazed *not* that the Romans did not provide a more precise timepiece, but that under their reckoning of hours they were able to provide an instrument that served daily needs at all. It required a hefty amount of ingenuity, but the Romans made their water clocks indicate the shifting length of hours from month to month, rather than from day to day. (The day-to-day changes were too small to be of practical interest.)

The equal hour did not arise until about the fourteenth century. Around 1330 the hour became our modern hour, one of twenty-four equal parts of a day. This new "day" included the night, and it was measured by the *average* time between one noon and the next, the average being over one year. For the first time in history, an "hour" took on a precise, year-round meaning everywhere.
This movement from the seasonal or "temporary" hour to the equal hour is a subtle but profound revolution in human experience. Here was humanity's declaration of independence from the sun, new proof of our growing mastery over ourselves and our surroundings. Only later would it be revealed that we had accomplished this mastery by putting ourselves under the dominion of a machine, with impervious demands all its own.

Procedure

Part A: The Solar and Sidereal Day

Why does the sun appear to move in our sky at all? The earth is spinning about an axis once a day *and* revolving around the sun once a year.

The rotation of the earth about its spin axis once a day has the effect that celestial objects seem to spin around the earth once a day—apart from any motion the celestial objects might have relative to the earth. If the earth did not revolve around the sun but simply spun on its axis, the sun would appear to go around the earth once a day—once a sidereal day. But the earth does revolve around the sun and this relative motion introduces a small complication: the time from one noon to the next (one solar day) is not the same as the time for the earth to spin once upon its axis (one sidereal day). This difference is so important that we will examine it in two different ways so that its origin is clear.

First, we will adopt a geocentric perspective, that is, we will consider the earth to be motionless and the celestial sphere and the objects on it to revolve around us. For concreteness, let us pick a fixed point on the celestial sphere, say the vernal equinox (see fig. 48.1), the intersection of the celestial equator and the ecliptic where the sun goes from south of the equator to north of it. (The sun in its motion along the ecliptic is at the vernal equinox on or about March 21.) Let us further suppose that it is about noon on March 21 so that the sun as well as the vernal equinox are on the local celestial meridian as in fig. 48.1a. The celestial sphere rotates (due to, of course, the earth rotating about its spin axis). Sometime during the next day that fixed point on the celestial sphere—the intersection of the ecliptic and celestial equator, the vernal equinox—is again on the meridian. That interval of time is one *sidereal* day. A fixed point on the celestial sphere has gone around once. But in that time the sun has moved eastward along the ecliptic as shown in fig. 48.1b. In fig. 48.1 b, it is not quite noon since the sun is not yet on the meridian. We have to wait a bit longer for the celestial sphere to continue to spin to bring the sun up to the meridian.

1. The sun moves 360° around the ecliptic in 365 days, so the sun moves about 1° per day along the ecliptic. Therefore, the sun is about 1° east of the vernal equinox. Calculate how long you have to wait after the situation depicted in fig. 48.1b so that the sun is on the meridian. Express your answer in minutes.

We have described one way of showing that a solar day is a bit longer than a sidereal day. Now we shall examine the situation from a different perspective, a heliocentric one.

In fig. 48.2, the earth is initially at position A in its orbit around the sun. It is noon for an observer at the foot of the dotted arrow. Consequently, it is midnight for an observer at the foot of the solid arrow at A. One sidereal day later, the earth is at B and the arrows have the same orientation as at A, but at B the daylight dotted arrow does not point to the sun and so it is not noon. A little later, the earth has turned more upon its axis and moved a bit in its orbit and now the dotted daylight arrow points toward the sun. The time from A to B constituted one sidereal day, while the time from A to C constituted one solar day.

2. Figure 48.3 depicts the sun's apparent motion along the ecliptic (where the zodiacal constellations are found) in a heliocentric perspective. As the earth orbits the sun the line of sight of the sun and toward the background stars also moves. For example, the diagram shows the sun in Leo. A month later the earth will have moved enough so that Virgo lies behind the sun.

Figure 48.1 (a) A day in late March with the Sun and vernal equinox on the local celestial meridian. (b) One sidereal day later Earth has rotated once on its axis and the vernal equinox is back on the celestial meridian. However, the Sun has moved eastward along the ecliptic.

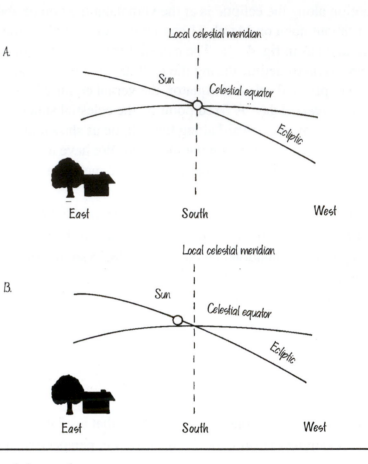

Figure 48.2 The difference between a sidereal day and a solar day arises from the rotation of the Earth about its axis and its revolution around the Sun. The illustration is a view of the Earth and Sun from far above the north pole of Earth.

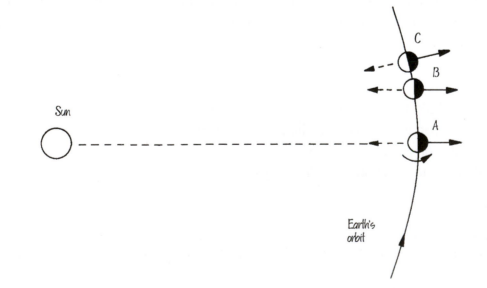

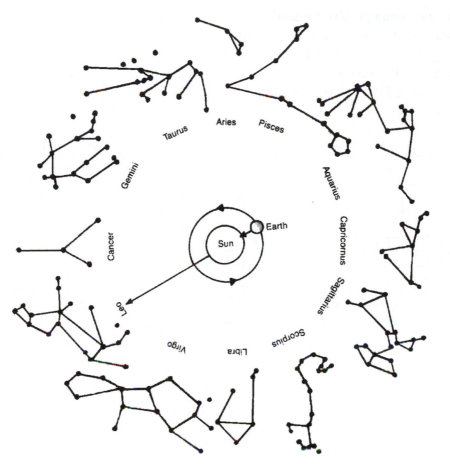

Figure 48.3

Part B: The Analemma—Details of the Sun's Motion

If the earth's orbit were circular and if the earth's spin axis were perpendicular to the plane of its orbit, the sun would always rise precisely in the east, move along the celestial equator at a constant rate and set precisely in the west. The sun would also appear to move eastward among the stars at a constant rate, completing one revolution in one year. In this idealized situation, the sun would be a perfect clock and would arrive on the observer's meridian at exactly equal intervals.

As you know, the earth's orbit around the sun is elliptical and the earth's spin axis is tilted 23 1/2° from the perpendicular. These circumstances, as you will see, will cause the time interval between successive meridian crossings of the sun to *vary* throughout the year. We can still refer to a fictitious *mean sun* that moves uniformly along the celestial equator and is on the meridian at noon and again precisely 24 hours later. The real sun, unfortunately, does not behave this way, but the corrections to the time kept by the real sun are not large. They can be represented on a three-coordinate plot called the **analemma** (L., sundial).

The analemma is a closed curve resembling a flatbottomed figure 8 (see fig. 48.4). You may have seen the analemma on a terrestrial globe where it is usually placed in the empty part of the Pacific

373

Figure 48.4 The analemma graphs the sun's
declination and the daily difference between clock
noon and noon by the sundial (sun on meridian for
every day of the year. Declination is the distance in
degrees north or south of the celestial equator.

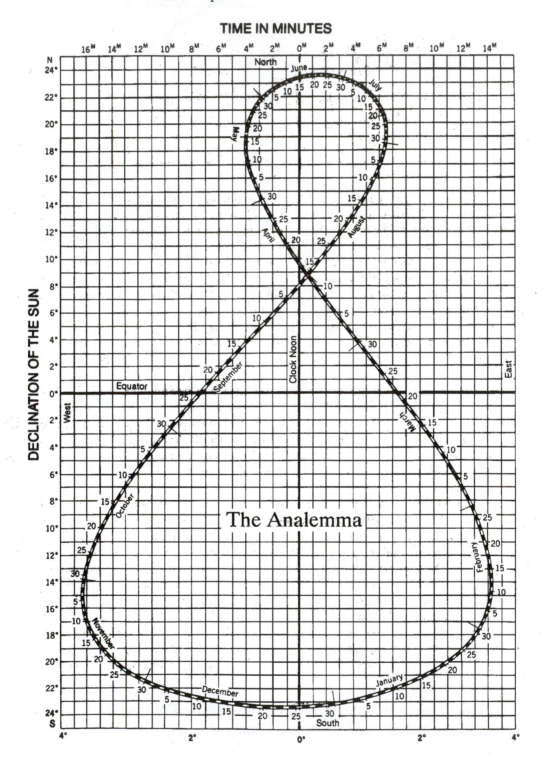

Figure 48.5 Illustration of how the equator-ecliptic angle affects the sun's timekeeping. At the equinox, E represents the solar motion along the ecliptic; its eastward component E' on the equator is shorter. At the solstice, S (equal to E) runs due eastward and the hour circles are closer together, the component S' is longer than both E' and E

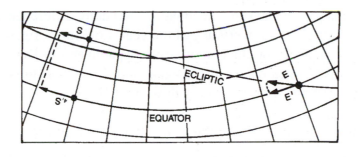

Ocean. Each point on the analemma represents a date in the year. The north-south coordinate at the point gives the sun's declination on that date. The east-west coordinate indicates the number of degrees (or minutes of time) by which the sun is east or west of the observer's meridian when the local mean solar time is noon. In the following, we shall examine the origin of the difference between the mean sun and the actual sun, and the information contained in the analemma.

The spin axis of the earth is inclined 23 1/2° to the plane of its orbit around the sun. Because of this tilt, the yearly path of the sun eastward among the stars (the ecliptic) is tilted 23 1/2° with respect to the celestial equator. In late June the sun is 23 1/2° north of the equator and in late December 23 1/2° south of it. This annual north-south oscillation of the sun's declination is responsible for the lengthwise extension of the analemma pattern.

The ecliptic tilt has yet another effect on the sun's motion. Since the sun moves along the ecliptic, which is tilted with respect to the celestial equator, the sun's motion relative to the stars is due east only in late June and late December. Hence, the sun's eastward advance per day is greatest at those times and least in March and September when the ecliptic crosses the equator at a slant (see fig. 48.5).

Because the meridians of right ascension are more closely spaced at declinations of ±23 1/2° than at the equator, the actual sun's effective eastward motion is faster than that of the mean sun's. Twice a year, near the solstices, the sun arrives later and later on the local meridian because of its relatively fast eastward motion from day to day (look again at fig. 48.1 a and b), and as a clock it runs slow. Twice a year near the equinoxes, the sun arrives on the observer's meridian earlier and earlier each day, and as a clock it runs fast. Therefore, two times during the year the actual sun is ahead of clock time and two times during the year it is behind. This effect gives rise to the east-west spread to analemma and determines its general figure-8 shape.

One further influence on the shape of the analemma arises because of the *elliptical* orbit of the earth around the sun. As Johannes Kepler discovered nearly four centuries ago, a planet moves fastest in its orbit near perihelion (point nearest the sun) and slowest at aphelion (point farthest from the sun). Since the earth reaches perihelion on January 3 and aphelion on July 7, the motion of the sun along the ecliptic is faster than average during the winter months and slower than average during the summer months. On January 3, the apparent rate of the sun along the ecliptic is 1.019 degrees per day, while on July 7 the sun moves at a rate of 0.953 degrees per day. The principal effect of this annual velocity variation of the sun is to broaden the southern loop of the analemma and compress the northern loop.

In summary, the analemma graphs the sun's declination, and daily difference between clock noon and noon by the sun (sun on meridian) for every day of the year. Looking at fig. 4.3 we see that the sun is west of the mean sun, that is, ahead of clock noon, from September 1 to December 26, falls behind from December 26 to April 15, then moves ahead again until June 15. It falls behind again until September 1, alternately speeding up and slowing down with respect to clock time.

1. Briefly describe the effect on the shape of the analemma if the ecliptic-equator angle were to increase.

2. Obtain from your instructor the latitude of your location and fill it in below.

latitude of your location = _____

Determine the altitude over the southern horizon of the intersection of the celestial equator and the local celestial meridian.

3. Use the analemma in fig. 48.4 and your answer to step 4 to fill in Table 48.1. The latitude of Altengaard, Norway is +70°.

Table 47.1 Declination and Noon Altitude of Sun			
Date	Declination of Sun	Noon Altitude of Sun at Your Location	Noon Altitude of Sun at Altengaard
February 15			
June 21			
October 15			
December 22			

4. Rip van Winkle awakens from his extended slumber and asks a passerby what year it is so he can determine how long he slept. The passerby quickly responds and then rapidly moves away. Realizing a twenty year snooze might be a world's record, he thinks he should pin down the date as well as the year. He moves out from under the tree where he slept and measures the altitude of the sun when it crosses the meridian; he finds it is 44° over the southern horizon. He knows that the latitude of his chosen spot on the Hudson River in New York is 42°. On what two possible dates could Rip van Winkle have awakened? He notices that buds are appearing on the tree he slept under. What is the date of his awakening?

As we saw previously, the analemma graphs the daily difference between clock noon and apparent noon when the sun crosses the meridian. As an example, check the analemma in fig. 4.3 to see that on October 15 the sun will cross the meridian 14 minutes before clock noon.

5. For the dates given in the table below, use the analemma in fig. 48.3 to determine whether the actual sun will cross the meridian before or after clock noon and by how many minutes.

Table 47.2 Sun Time and Clock Time		
Date	Sun Crosses Meridian Before or After Clock Noon	Amount of Time Before or After Clock Noon
January 10		
February 15		
April 15		
July 30		
September 1		
October 30		

6. Determine the altitude over the southern horizon of the actual sun and the time when it crosses your local meridian on the dates in the table below.

Table 47.3 Sun Altitude and Time of Meridian Crossing		
Date	Altitude of Sun	Time When Sun Crosses Meridian
May 15		
June 15		
November 4		
December 25		

Experiment 49: Phases of Moon

Introduction

Around the world and in every era, people have scrutinized the moon. Its influence has seeped into our language, where we find relics of mythic, mystic, and romantic meanings—in such words as "moonstruck" and "lunatic" (Latin *Luna* means moon), in "moonshine," and in the moonlit setting of lovers' meetings. Even deeper is the primeval connection between the moon and measurement. The word "moon" in English and its cognates in other languages are rooted in the base *me* meaning measure, as in the Greek *metron*, and in the English *meter* and *measure*, reminding us of the moon's service as the first universal measure of time. The ancient German communities, Tacitus reported nearly two thousand years ago, held their meetings at new or full moon, "the seasons most auspicious for beginning business."

The phases of the moon were convenient worldwide cycles that anybody could see and so were the basis for reckoning time in the construction of a calendar. The Babylonians and ancient Israel, Greece, and Rome all used lunar calendars. The Muslim world, with a literal obedience to the words of the prophet Muhammad and to the dictates of the holy Koran, continues to live by the cycles of the moon.

Despite or because of its easy use as a measure of time, the moon proved to be a trap for naive mankind. For while the phases of the moon formed a convenient cycle, they were an attractive dead end. What hunters and farmers most needed was a calendar based on the seasons—a way to predict the coming of rain or snow, of heat and cold. How long until planting time? When to expect the first frost? The heavy rains?

For these needs the moon gave little help. The seasons of the year, as you know, are governed by the movements of the earth around the sun or equivalently, the movement of the sun against the background stars. The discomforting fact that the cycles of the moon and the cycles of the sun are incommensurate would stimulate thinking, however. Had it been possible to calculate the year, the cycle of seasons, simply by multiplying the cycles of the moon, humans would have been saved a lot of trouble. But we might also have lacked the incentive to study the heavens and to become astronomers.

Using about six observations of the moon over a two-week interval, you will examine how the moon changes position in the sky and how its appearance changes as it moves.

Procedure

Part A: The Observations

A complete cycle of lunar phases requires about 29 1/2 days. You will follow the moon for at least half a cycle so your observations will be spread over the course of about two weeks. During this time you will be recording both the appearance of the moon, that is, how much of the lunar disk is illuminated, and also the *position* of the moon relative to the landscape and the background stars.

A. Choosing a Location for Your Observations

You will need to find a spot with a relatively unobstructed view of the horizon and the sky. The middle of a large open space with a few low-profile landmarks near the horizon is ideal. You will need a little ambient light to make some sketches to record your observations, but you should not be near any bright sources of light. *It is important that you use precisely the same location for all your observations*, so if you make them from the middle of an open field or a parking lot, mark the location of your observing spot so that you can return to it.

B. Timing Your Observations

Your instructor will give you some general guidelines as to dates and times for when to begin your series of observations, but you can choose a time of night convenient to you for observing the moon. Once you choose a time of night for observing, *it is critical that you make all your observations at the same time of night to within 15 minutes.*

Because the weather may hamper your observations over the course of the observing period, you do not want to let many clear nights go by. You should try to make observations every other night, weather permitting. After your first observation, which might require about half an hour, succeeding observations of the moon will require much less time, about five minutes.

C. Recording the Observations

In order to record the appearance and position of the moon, you will need to make a scale drawing of the moon in its relation to the landmarks around your observing site. This sketch can be made on a standard poster board (22" × 28") or a similarly sized sheet of drawing paper. You may want to initially go out about a half hour before the optimum observing time that you have chosen so that you can get enough information to complete the sketch prior to your first lunar observation.

To make a scale drawing, you will need to measure the height, width, and separations of a few landmarks around your observing site. One way is to make measurements in degrees. Convenient standards are the following:

1. the width of your outstretched hand held at arms length is about 22°
2. the width of your fist held at arms length (knuckle to knuckle) is about 8°
3. the width of your three middle fingers held at arms length is about 5°
4. the width of your index finger at arms length is about 2°

A convenient scale to use would be 5° in the sky and landscape corresponding to 2 centimeters on the drawing.

Alternately you can carry a ruler with you and measure the dimensions and separations of landmarks with the ruler held at arms length. In this case, a convenient scale to use is 2 centimeters on the ruler corresponding to I centimeter on the drawing.

To plot the position of the moon on your drawing, you will need to make two measurements. Choose a convenient landmark near the moon and from a particular point on the landmark measure *how far over* and *how far up* the moon's center is located.

**Estimate the angular size of the moon. (The width of your index finger at arms length near the top knuckle is about 2°.)

Record the appearance of the moon at the plotted position. You may want to forget about getting the moon's size to scale and use any conveniently sized object like a coin to draw in the moon and indicate the degree to which it is illuminated.

Record the positions of a few bright stars near the moon on your first observing session. Number your observations of the moon and in a corner of your drawing record the date and time to the nearest minute of each observation.

Determine the direction of one or two compass points (e.g., west and south) and indicate them on your drawing.

1. How would you describe the night-to-night path of the moon with respect to the landmarks around your observing site?

2. In which direction does the moon move with respect to the background stars from night to night?

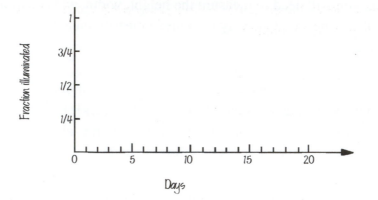

On the graph above, plot the fraction of the lunar disk illuminated as a function of the number of days after your first observation. (Your first observation is at 0 days.)

For each of the moon's positions I through 8 in fig. 7.1, shade the dark portion of the moon that would be seen for an observer high above the plane of the earth-moon system.

In fig. 7.2, shade in below what an earthbound observer would see for the corresponding positions.

Part B: Lunar Phase Nomenclature

The explanation of the phases is quite simple: the moon is a sphere and at all times the side that faces the sun is illuminated and the side that faces away from the sun is dark. From the previous drawings, you can see that the phase of the moon that we see from earth, as the moon revolves around us, depends on the relative orientation of the sun, moon, and earth.

When the moon is almost exactly between the earth and sun, the dark side of the moon faces us. We call this "new moon." A few days earlier or later we see a sliver of the lighted side of the moon and call this a "crescent." As the month wears on, the crescent gets bigger until half the disk of the moon is illuminated. We sometimes call this "half moon." However, since this occurs one-fourth of the way through the cycle of phases, the situation is also called a "first quarter moon."

When over half of the moon's disk is visible, we have a "gibbous" phase. When the moon is on the opposite side of the earth from the sun, the entire face of the moon is illuminated; this is a "full moon." A few days later the moon is noticeably in a gibbous phase again and soon thereafter we see a half moon again, called third quarter this time. Then we go back to a crescent phase and "new moon" again. The cycle then repeats.

When the amount of illuminated moon is increasing from day to day, the moon is said to be "waxing." So from new moon to full we have waxing crescent and gibbous phases. When it is decreasing, after full moon until the next new moon, the phases are said to be "waning."

1. In the spaces below, write down the names of the phases corresponding to sketches in fig. 49.2. Include whether a phase is waxing or waning.

1. _____   5. _____
2. _____   6. _____
3. _____   7. _____
4. _____   8. _____

Positions a, b, c, and d on fig.49.1 correspond to observers looking at different times of the day. Observer a is observing halfway between sunrise and sunset, i.e., at noon. Observer b is on the boundary between day and night and so is observing at sunset. (Why not sunrise?) Observer c is observing at midnight, while d is at sunrise.

2. Place a cross on the earth representing your observing time.

Figure 49.1 The earth-moon system as viewed from far above the earth's north pole with different moon positions 1-8.

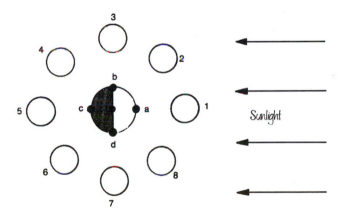

Figure 49.2 Lunar phases seen by an earthbound observer.

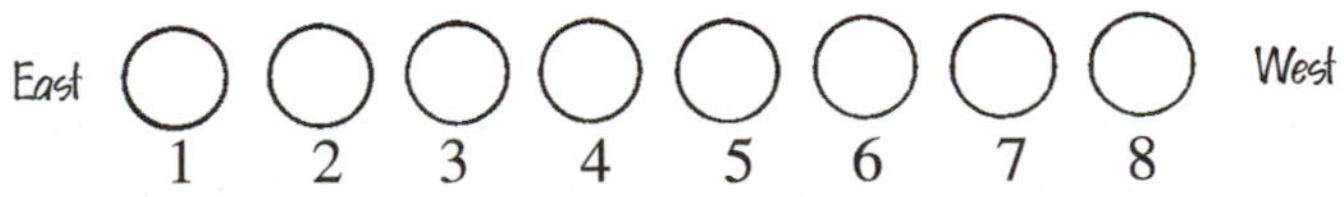

3. At what time of the day do you expect:

(a) the full moon to rise?

(b) the new moon (or a very thin crescent moon) to rise?

(c) the first quarter moon to rise?

(d) the third quarter moon to rise?

(e) the first quarter moon to set?

(f) the new moon (or a very thin crescent moon) to set?

Experiment 50: Power Output of Sun

Introduction

The sun powers the earth and the life that flourishes upon it. It warms us and feeds us, and illuminates our world. Life on earth is intimately tethered to the energy radiated by the sun. Birds greet the sunrise with audible ecstasy, and even some one-celled organisms know to swim to the light. Think of the sun's heat on your upturned face on a cloudless summer's day; think how dangerous it is to gaze at the sun directly. From 150 million kilometers away, we recognize its power. It is powerful beyond human experience. Our ancestors worshiped the sun, and they were far from foolish.

For all its furious energy output, the sun is a rather reliable source of energy. Some of the best measurements of the rate of energy emission from the sun were made by instruments aboard the spacecraft called the Solar Maximum Mission. From above the earth's atmosphere, it was able to measure the sun's rate of energy emission with high precision, and it found variations of about 0.1 percent over the course of a few weeks. These variations appear to be related to changing numbers of sunspots. Such short-term variations do not seem to have much of an effect on the earth or its weather, but a 0.1 percent variation that lasted for a decade might have a very significant impact, especially on world agriculture. A variation of only I percent would change the average temperature of the earth by 1-2°C (about 1.8-3.6°F). For comparison, during the last ice age, the average temperature on the earth was about 5°C cooler than it is now.

What could possibly be the source of the sun's energy that could have sustained its prodigious energy output for so long? During the late nineteenth century, when the earth was found to be billions of years old, the source of the sun's power became an embarrassingly difficult puzzle to explain. The crux of the problem was how to explain both the rate of energy production and its longevity. The sun has been roughly the same luminosity for at least 3 billion years. How do we know? Geologists have found rocks containing fossils of living organisms that are at least that old. For life to exist, the earth must be warm, and, therefore, the sun must have been roughly at its present luminosity.

What about chemical reactions as the source of the sun's energy? Energy from ordinary chemical reactions, such as burning, could not explain both the energy production rate and the time scale over which that rate must be maintained. If the sun were composed entirely of oxygen and coal (carbon), to maintain the observed energy output, the one you will measure, the sun would have burned to a dark cinder in about 20,000 years. Obviously, the sun could not be a coal-burning furnace!

In the middle of the nineteenth century Herman von Helmholtz (1821-1890) and William Thomson Lord Kelvin (1842-1907) proposed that the sun shone because it was releasing gravitational energy by shrinking, that is, gravitational contraction converted gravitational potential energy to radiative energy. As the sun contracts, the gases get squeezed, with the result that the temperature rises with a concomitant emission of radiation.

Because of the sun's substantial mass, a contraction rate of only 40 m/year would liberate the required energy. The gravitational energy stored in the sun would last for about 20 million years, far longer than the age of the earth as determined by early geologists.

1. The sun has a diameter of 1.4×10^9 m and its distance is 1.5×10^{11} m. (a) Verify that the *angular* diameter of the sun is 0.53°. (b) Assuming that gravitational contraction as described above is responsible for the energy output of the sun, by how much would angular diameter of the sun have decreased over a thousand years? Typical accuracies for an angular measurement in ancient times (~2,000 years ago) was a few minutes of arc (~5). Could we have noticed a decrease in the size of the sun from an ancient measurement?

Near the beginning of the twentieth century, the study of nuclear transformations indicated that the energies involved in the atomic nucleus were a million times larger than the energies involved in chemical reactions, which were due to the rearrangement of the outer electrons among atoms. Here was the possibility of a tremendous energy source.

Albert Einstein provided the key idea about the sun's energy. Grappling with the fundamental nature of electromagnetic waves, he demonstrated that mass and energy are related by the equation $E = mc^2$, where E is the energy (in joules) released in the conversion of a mass m (in kilograms) and c is the speed of light (in meters per second).

But how to change matter into energy?

Not until the 1930's did the mechanism by which the sun generated its energy from nuclear reaction become understood. The energy comes from the same reactions that power a hydrogen bomb. They are called **fusion reactions** because they fuse nuclei together.

A number of different fusion reactions generate the sun's energy, but one series of reactions having the net effect of fusing four hydrogen nuclei to make a single helium nucleus accounts for more than 85 percent of the energy produced by the sun. One helium nucleus has 0.7 percent less mass than four hydrogen nuclei, and most of the mass difference appears as energy.

4 hydrogen nuclei	.... 6.693×10^{-27} kg
1 helium nucleus	.... 6.645×10^{-27} kg
difference in mass	.... 0.048×10^{-27} kg

Using Einstein's equation

$$E = mc^2$$

$$= (0.048 \times 10^{-27} \text{ kg})(3 \times 10^8 \text{ m/s})^2$$

$$= 4.3 \times 10^{-12} \text{ Joules}$$

In this experiment you will measure the power output of the sun. Your answer will be expressed in units of watts (energy/time) or, equivalently, in Joules/s. Once you know the power output of the sun in Joules/s, you can determine the number of reactions per second needed to maintain the observed power level. Finally, you will be able to estimate the lifetime of the sun from your observations.

Procedure

Imagine a source of light, say a 100-W bulb. The 100 W means that the bulb emits 100 Joules of energy each second (1 W = 1 J/s). Now imagine this bulb at the center of a sphere 2 m in radius. The energy emitted by the bulb in the form of electromagnetic radiation is spread out over the surface of the sphere. The intensity of the light, the power per unit area, is

$$I = \frac{P}{A} = \frac{100 \text{ W}}{4\pi(2\text{m})^2} = 2 \text{ W} / \text{m}^2$$

If we imagined another sphere surrounding the light bulb with a radius of 4 m (twice 2 m), the intensity of the light on this surface would be

$$I = \frac{P}{A} = \frac{100 \text{ W}}{4\pi(4\text{m})^2} = \frac{1}{2} \text{ W} / \text{m}^2$$

one-fourth what it was at 2 m. We can generalize these examples by noting that the intensity of light falls off as the square of the distance from the source.

In this experiment, you will use a null photometer to balance the power levels of sunlight and the light from a standard light bulb.

The null photometer is a box open at both ends that allows the light intensity from two different sources to be compared and balanced.

At the center of the null photometer are two blocks of paraffin cemented to sheets of aluminum foil. The light from some source enters the photometer and gets scattered inside the paraffin. Some of the light gets scattered to the side of the paraffin block where it can be viewed. The aluminum foil

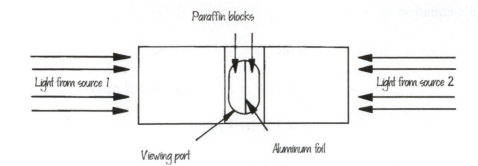

Paraffin blocks

Light from source 1

Light from source 2

Viewing port

Aluminum foil

Figure 50.1

prevents light from one source from leaking over to the other side of the photometer where it would interfere with the light level measurement of the second source. The aluminum foil also reflects light from the source, giving a greater efficiency to the scattering process.

In making a measurement with the null photometer, the relative positions along a line of the two sources and photometer are adjusted so that the scattered light intensity from both paraffin blocks are equal as seen from the viewing port.

The human eye/brain detection system is quite good at matching light levels and perceiving small differences in light intensity. When color differences are present between the light from the two sources, the balance is more difficult to achieve and reproduce. To remove to a degree the effects of color differences, a piece of colored plastic is placed over the viewing part so that the eye is balancing the same color.

Part A: A Controlled Experiment

To verify the experimental technique we will use to measure the power output of the sun, we will first perform a controlled experiment in the lab. Arrange two light bulbs of differing power outputs along a line, as shown below, with the null photometer placed between them.

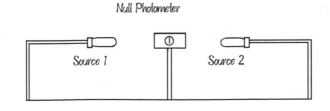

Null Photometer

Source 1

Source 2

Figure 50.2

Be careful to place all the components at the same height and along a line.

Adjust the relative positions of the sources and photometer so that the same intensity is registered on both halves of the photometer's viewing port. Measure the relative distances of source 1 to the photometer and from source 2 to the photometer. Measure from the center of the bulbs spherical portion to the center of the null photometer.

When the null photometer is balanced, the intensity of light of source 1 is the same as the intensity of source 2.

$$\frac{P_1}{4\pi(d_1)^2} = \frac{P_2}{4\pi(d_2)^2}$$

2. Using one source as a standard, and the relative distances you measured, predict what the power output of the second source should be. Check to see whether the rating on the second bulb corresponds to your prediction.

Part B: Diffusion

The intensity of sunlight is not easy to match with a light bulb, but the null photometer can be modified so that it is able to compare two different light sources with very large differences in power output.

The intensity of a very bright light source can be reduced by stopping down the end of the photometer pointing to the bright source. The reduction can be achieved by diminishing the size of the opening from which light can reach the paraffin.

Set up the two sources and photometer as in Part I of this experiment, except on the side of the photometer toward the brighter source attach the mask that lowers the effective area of the null photometer when compared to the area A_2 of the other side of the photometer. The side toward the brighter source will be measuring a power output diminished by a factor of A_1/A_2, where A_1 is the area over which light is admitted into the stopped-down portion of the photometer.

[A diffuser is mounted over the hole in the mask. By diffusing the light reaching the paraffin, a more uniform illumination is achieved.]

3. Verify that the modified null photometer is still able to predict the value of the power output of the fainter source.

Part C: Power Output of Sun

Take the null photometer with the aperture stop and the brighter light bulb outside and measure the power output of the sun. (The distance to the sun is 1.5×10^{11} m.)

Power output of the sun = _____

4. From the introduction, it was seen that one fusion reaction produces 4.3×10^{-12} J. How many reactions each second are required to sustain the power output of the sun that you measured?

5. How much mass is the sun losing each second in making this mass-to-energy conversion?

6. The sun's mass is 2×10^{30} kg and is 75% by mass hydrogen. Because the nuclear reactions occur only near the hot, dense core, only about 15% of the total amount of hydrogen is available for fusion reactions. Estimate the total energy available for radiation. (There are 6×10^{26} hydrogen nuclei in 1 kg of hydrogen.)

7. From the total energy available, estimate the lifetime of the sun assuming the sun has been radiating the power level you measured over its entire lifetime.

Experiment 51: Special Project

The **Special Project** (SP) is an independent investigation that is accomplished outside of the classroom and laboratory room, with ordinary and everyday devices. *No laboratory equipment* will be available for this project since the SP is intended to be an unstructured "kitchen science" investigation. You will need to take this your-stuff-only restriction into consideration when deciding what you are going to investigate as you *must* use only available equipment to do the investigation. This restriction is an important part of the process. The SP is an opportunity to explore that area of physical science that interests you most, and in a real world situation. You are free to pursue any physical science concept for your experiment, but your project should not simply mimic one of the regular experiments assigned in class or one from the laboratory manual. As best you are able to, you are to play the role of an original, creative thinker during this investigation.

Important: A SP proposal must be submitted to your laboratory instructor, and *approved* before beginning any work. This is necessary for safety considerations, to avoid project duplications, and to ensure that your project satisfies the intent of the assignment. Please read the guidelines below, and also read the evaluation sheet to make sure you understand the intent of the SP. You will be informed of the due date for the proposal, which should briefly include the following:

- Name(s) of the person(s) working as an individual or as a team.
- A brief statement of the question to be answered.
- What equipment will be used in what procedure to find the answer.
- How measurements will be made; how data will be collected.

You will have about 30 days to conduct the SP outside of class. One lab period has been established as a reporting session, a time when the experimental findings will be presented in both oral and written formats. The oral presentation will not be as detailed as the written one, however, it is the only way the rest of the class will be able to benefit from your efforts. It therefore warrants some thought and creativity.

The SP **oral presentation** should include the following as possible:

- Reporting responsibilities should be equally divided among all team members.
- Describe the question to be answered and how the investigation was conducted.
- The independent and dependent variables should be identified.
- Use display boards to show diagrams, data tables, and graphs.
- An interpretation of results using graphic analysis (mathematical model).
- Explanations for the deviation (if any) of results from what was expected.
- Concluding statements.

The SP **written report** should be typed following the outline below. An appropriate title for your project should be in the center of the first page with the names of each team member in the upper right hand corner. Each of the following sections should be included in your report prefaced with the appropriate heading.

Purpose: A brief statement of the question that was investigated.

Apparatus: A diagram of the equipment with all parts labeled, showing the experimental setup.

Procedure: The sequence for conducting the experiment stated in brief sentences. The independent and dependent variables should also be clearly identified, including a short statement of how the independent variables were controlled.

Raw Data: The values measured directly from the experiment with data from as many trials as judged necessary (a minimum of three trials is required). Data should be organized into neat tables with the units (m, kg, s, etc.) clearly labeled. This section should be the original handwritten data sheet written at the time of the experiment.

Evaluation Of Data: A presentation of findings via graphs and formal data tables. Formal data tables include averaged values of your multiple trials as well as processed data in a spreadsheet format. Processed data refers to calculated values derived at by inserting experimental values into various equations. State what equations were used and identify the symbols used in the equations. If repetitive calculations are performed, show only one example calculation. All other calculated values will appear in your formal data tables. Graphs should be labeled with the variables on the appropriate axes and units indicated clearly. Interpret your graphs with statements of relationships between the variables. These statements need to be complete English sentences. A mathematical model for the graph should be found (if possible) and stated in this section also. An equation for a line with the slope and *y*-intercept given in the proper units is an example of such a mathematical model. This section also should contain a statement describing the quality of the results.

Conclusion: Results are compared to what was expected and plausible explanations offered for any deviation. The meaning of the slope of a graph and any equations derived from graphical analysis are also stated here. You should also state whether the goals of your experiment were accomplished or not.

Note: *As the due date for the SP proposal and the presentation day approaches please feel free to contact your laboratory instructor for individual help, advice, or encouragement.*

SP Evaluation Scorecard

Team Members:

Proposal (5 points)---_____

    Turned in on time and clearly stated.

Format of Report (4 points)---_____

    1. Group names, title.
    2. Each section of report clearly labeled, neat, and organized.

Purpose of Investigation (2 points)---_____

    Question to be answered by experiment is clearly identified and stated.

Procedure (6 points)---_____

    1. Independent and dependent variables are clearly defined and controlled.
    2. Clear, brief sequence of steps.
    3. Diagram(s) drawn with all components labeled.

Raw Data (6 points)---_____

    1. Measurement data organized into neat tables.
    2. Values are clearly labeled.
    3. Multiple trials.

Evaluation Of Data (12 points)---_____

    1. Tables and/or sample calculations.
    2. Graphs; variables on appropriate axes, use of units.
    3. Interpretation of graphs
        a. Written statement of relationship.
        b. Mathematical model (equation, units on slope).

Conclusion (12 points)---_____

    1. Written explanation (English sentences) of relationships.
    2. Meaning of slope in terms of experimental question.
    3. General equation included.
    4. Reasonable explanation for divergent results (when applicable).

Presentation (3 points)--_____

    1. Display clear and understandable.
    2. Team functioned well together.
    3. Team seemed knowledgeable in their presentation.

Total Points--_____

Appendix I: The Simple Line Graph

An equation describes a relationship between variables, and a graph helps you "picture" this relationship. A line graph pictures how changes in one variable go with changes in a second variable; that is, how the two variables change together. One variable usually can be easily manipulated; the other variable is caused to change in value by manipulation of the first variable. The *manipulated variable* is known by various names (*independent, input, or cause variable*) and the *responding variable* is known by various related names (*dependent, output, or effect variable*). The manipulated variable is usually placed on the horizontal or *x*-axis of the graph, so you can also identify it as the *x-variable*. The responding variable is placed on the vertical or *y*-axis. This variable is identified as the *y-variable*.

The graph in appendix figure I.1 shows the mass of different volumes of water at room temperature. Volume is placed on the *x*-axis because the volume of water is easily manipulated and the mass values change as a consequence of changing the values of volume. Note that both variables are named, and the measuring unit for each variable is identified on the graph.

The graph also shows a number scale on each axis that represents changes in the values of each variable. The scales are usually, but not always, linear. A *linear* scale has equal intervals that represent equal increases in the value of the variable. Thus a certain distance on the *x*-axis to the right represents a certain increase in the value of the *x*-variable. Likewise, certain distances up the *y*-axis represent certain increases in the value of the *y*-variable. In the example, each mark has a value of

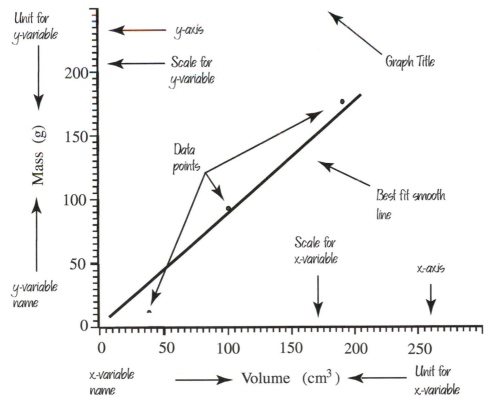

Appendix Figure I.1

395

five. Scales are usually chosen in such a way that the graph is large and easy to read. The *origin* is the only point where both the *x*- and *y*-variables have a value of zero at the same time.

The example graph has three data points. A *data point* represents measurements of two related variables that were made at the same time. For example, a volume of 190 cm^3 of water was found to have a mass of 175 g. Locate 190 cm^3 on the *x*-axis and imagine a line moving straight up from this point on the scale (each mark on the scale has a value of 5 cm^3). Now locate 175 g on the *y*-axis and imagine a line moving straight out from this point on the scale (again, note that each mark on this scale has a value of 5 g). Where the lines meet is the data point for the 190 cm^3 and 175 g measurements. A data point is usually indicated with a small dot or an x; a dot is used in the example graph.

A "best-fit" smooth, straight line is drawn as close to all the data points as possible. If it is not possible to draw the straight line *through* all the data points (and it usually never is), then a straight line should be drawn that has the same number of data points on both sides of the line. Such a line will represent a best approximation of the relationship between the two variables. The *origin* is also used as a data point in the example because a volume of zero will have a mass of zero. In any case, the dots are *never* connected as in dot-to-dot sketches. For most of the experiments in this lab manual a set of perfect, error-free data would produce a straight line. In such experiments it is not a straight line because of experimental error, and you are trying to eliminate the error by approximating what the relationship should be.

The smooth, straight line tells you how the two variables get larger together. If the scales on both the axes are the same, a 45° line means that the two variables are increasing in an exact direct proportion. A more flat or more upright line means that one variable is increasing faster than the other. The more you work with graphs, the easier it will become for you to analyze what the slope means.

Appendix II: The Slope of a Straight Line

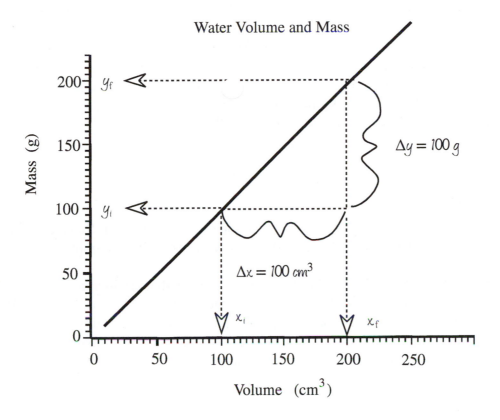

Appendix Figure II.1

   One way to determine the relationship between two variables that are graphed with a straight line is to calculate the **slope.** The slope is a ratio between the changes in one variable and the changes in the other. The ratio is between the changes in the value of the x-variable compared to the changes in the value of the y-variable. The symbol Δ (Greek letter Delta) means "change in," so the symbol Δx means "change in x." The first step in calculating the slope is to find out how much the x-variable is changing (Δx) in relation to how much the y-variable is changing (Δy). You can find this relationship by first drawing a dashed line to the right of the straight line so that the x-variable has increased by some convenient unit as shown in the example in appendix figure II.1. *Where you start or end this dashed line will not matter since the ratio between the variables will be the same everywhere on the graph line.* However, it is very important to remember when finding a slope of a graph to *avoid using data points* in your calculations. Two points whose coordinates are easy to find should be used instead of data points. One of the main reasons for plotting a graph and drawing a best-fit straight line is to smooth out any measurement errors made. Using data points directly in calculations defeats this purpose.

   The Δx is determined by subtracting the final value of the x-variable on the dashed line (x_f) from the initial value of the x-variable on the dashed line (x_i), or $\Delta x = x_f - x_i$. In the example graph above, the dashed line has a x_f of 200 cm³ and a x_i of 100 cm³, so Δx is 200 cm³ – 100 cm³, or 100 cm³. *Note that Δx has both a number value and a unit.*

397

Now you need to find Δy. The example graph shows a dashed line drawn back up to the graph line from the x-variable dashed line. The value of Δy is $y_f - y_i$. In the example, $\Delta y = 200$ g $-$ 100 g. The slope of a straight graph line is the ratio of Δy to Δx, or

$$\text{Slope} = \frac{\Delta y}{\Delta x}$$

In the example,

$$\text{Slope} = \frac{100 \text{ g}}{100 \text{ cm}^3}$$

or

$$\text{Slope} = 1 \text{ g}/\text{cm}^3$$

Thus the slope is 1 g/cm^3 and this tells you how the variables change together. Since g/cm^3 is also the definition of density, you have just calculated the density of water from a graph.

Note that the slope can be calculated only for two variables that are increasing together (variables that are in direct proportion and have a line that moves upward and to the right). If variables change in any other way, mathematical operations must be performed to *change the variables into this relationship*. Examples of such necessary changes include taking the inverse of one variable, squaring one variable, taking the inverse square, and so forth.

Appendix III: Experimental Error

All measurements are subject to some uncertainty, as a wide range of errors can and do happen. Measurements should be made with great accuracy and with careful thought about what you are doing to reduce the possibility of error. Here is a list of some of the possible sources of error to consider and avoid.

Improper Measurement Technique. Always use the smallest division or marking on the scale of the measuring instrument, then estimate the next interval between the shown markings. For example, the instrument illustrated in appendix figure III.1 shows a measurement of 2.45 units, and the .05 is estimated because the reading is about halfway between the marked divisions of 2.4 and 2.5. If you do not estimate the next smallest division you are losing information that may be important to the experiment you are conducting.

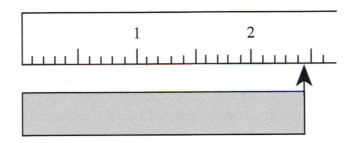

Appendix Figure III.1

Incorrect Reading. This is an error in reading (misreading) an instrument scale. Some graduated cylinders, for example, are calibrated with marks that represent 2.0 mL intervals. Believing that the marks represent 1.0 mL intervals will result in an incorrect reading. This category of errors also includes the misreading of a scale that often occurs when you are not paying sufficient attention to what you are doing.

Incorrect Recording. A personal mistake that occurs when the data are incorrectly recorded; for example, making a reading of 2.54 units and then recording a measurement of 2.45 units.

Assumptions About Variables. A personal mistake that occurs when there is a lack of clear, careful thinking about what you are doing. Examples are an assumption that water always boils at a temperature of 212° F (100° C), or assuming that the temperature of a container of tap water is the same now as it was 15 minutes ago.

Not Controlling Variables. This category of errors is closely related to the assumptions category but in this case means failing to recognize the influence of some variable on the outcome of an experiment. An example is the failure to recognize the role that air resistance might have in influencing the length of time that an object falls through the air.

Math Errors. This is a personal error that happens to everyone but penalizes only those who do not check their work and think about the results and what they mean. Math errors include not using significant figures for measurement calculations.

Accidental Blunders. Just like math errors, accidents do happen. However, the blunder can come from a poor attitude or frame of mind about the quality of work being done. In the laboratory, an example of a lack of quality work would be spilling a few drops of water during an experiment with an "Oh well, it doesn't matter" response.

Instrument Calibration. Errors can result from an incorrectly calibrated instrument, but these errors can be avoided by a quality work habit of checking the calibration of an instrument against a known standard, then adjusting the instrument as necessary.

Inconsistency. Errors from inconsistency are again closely associated with a lack of quality work habits. Such errors could result from a personal bias; that is, trying to "fit" the data to an expected outcome, or using a single measurement when a spread of values is possible.

Whatever the source of errors, it is important that you recognize the error, or errors, in an experiment and know the possible consequence and impact on the results. After all, how else will you know if two seemingly different values from the same experiment are acceptable as the "same" answer or which answer is correct? One way to express the impact of errors is to compare the results obtained from an experiment with the true or accepted value. Everyone knows that percent is a ratio that is calculated from

$$\frac{\text{Part}}{\text{Whole}} \times 100\% \text{ of whole} = \% \text{ of part}$$

This percent relationship is the basic form used to calculate a percent error or a percent difference.

The **percent error** is calculated from the *absolute difference* between the experimental value and the accepted value (the part) divided by the accepted value (the whole). Absolute difference is designated by the use of two vertical lines around the difference, so

$$\% \text{ Error} = \frac{|\text{Experimental value} - \text{Accepted value}|}{\text{Accepted value}} \times 100\%$$

Note that the absolute value for the part is obtained when the smaller value is subtracted from the larger. For example, suppose you experimentally determine the frequency of a tuning fork to be 511 Hz but the accepted value stamped on the fork is 522 Hz. Subtracting the smaller value from the larger, the percentage error is

$$\frac{|522 \text{ Hz} - 511 \text{ Hz}|}{522 \text{ Hz}} \times 100\% = 2.1\%$$

You should strive for the lowest percentage error possible, but some experiments will have a higher level of percentage errors than other experiments, depending on the nature of the measurements required. In some experiments the acceptable percentage error might be 5%, but other experiments could require a percentage error of no more than 2%.

A true, or accepted, value is not always known so it is sometimes impossible to calculate an actual error. However, it is possible in these situations to express the error in a measured quantity as a percent of the quantity itself. This is called a **percent difference**, or a percent deviation from the mean. This method is used to compare the accuracy of two or more measurements by seeing how consistent they are with each other. The percent difference is calculated from the *absolute difference* between one measurement and a second measurement, divided by the average of the two measurements. As before, absolute difference is designated by the use of two vertical lines around the difference, and

$$\% \text{ Difference} \ = \ \frac{|\text{One value} - \text{Another value}|}{\text{Average of the two values}} \ \times \ 100\%$$

Appendix IV: Significant Figures

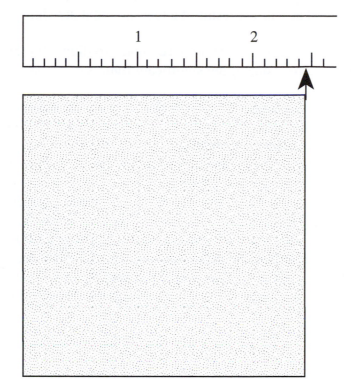

The numerical value of any measurement will always contain some uncertainty. Suppose, for example, that you are measuring one side of a square piece of paper as shown above. You could say that the paper is *about* 2.5 cm wide and you would be correct. This measurement, however, would be unsatisfactory for many purposes. It does not approach the true value of the length and contains too much uncertainty. It seems clear that the paper width is larger than 2.4 cm but shorter than 2.5 cm. But how much larger than 2.4 cm? You cannot be certain if the paper is 2.44, 2.45, or 2.46 cm wide. As your best estimate, you might say that the paper is 2.45 cm wide. Everyone would agree that you can be certain about the first two numbers (2.4) and they should be recorded. The last number (0.05) has been estimated and is not certain. The two certain numbers, together with one uncertain number, represent the greatest accuracy possible with the ruler being used. The paper is said to be 2.45 cm wide.

A **significant figure** is a number that is believed to be correct with some uncertainty only in the last digit. The value of the width of the paper (2.45 cm) represents three significant figures. As you can see, the number of significant figures can be determined by the degree of accuracy of the measuring instrument being used. But suppose you need to calculate the area of the paper. You would multiply 2.45 cm × 2.45 cm and the product for the area would be 6.0025 cm^2. This is a greater accuracy than you were able to obtain with your measuring instrument. *The result of a calculation can be no more accurate than the values being treated.* Because the measurement had only three significant figures (two certain, one uncertain), then the answer can have only three significant figures. Thus the area is correctly expressed as 6.00 cm^2.

403

There are a few simple rules that will help you determine how many significant figures are contained in a reported measurement.

Rule 1. All digits reported as a direct result of a measurement are significant.

Rule 2. Zero is significant when it occurs between nonzero digits. For example, 607 has three significant figures and the zero is one of the significant figures.

Rule 3. In figures reported as *larger than the digit one,* the digit zero is *not significant* when it follows a nonzero digit to indicate place. For example, in a report that "23,000 people attended the rock concert," the digits 2 and 3 are significant but the zeros are not significant. In this situation the 23 is the measured part of the figure and the three zeros tell you an estimate of how many attended the concert, that is, 23 thousand. If the figure is a measurement rather than an estimate, then it is written *with a decimal point after the last zero* to indicate that the zeros *are* significant. Thus 23,000 has *two* significant figures (2 and 3), but 23,000. has *five* significant figures. The figure 23,000 means "about 23 thousand" but 23,000. means 23,000. and not 22,999 or 23,001. One way to show the number of significant figures is to use scientific notation, e.g., 2.3×10^3 has two significant figures, 2.30×10^3 has three, and 2.300×10^4 has four significant figures. Another way to show the number of significant figures is to put a bar over the top of a significant zero if it could be mistaken for a place-holder.

Rule 4. In figures reported as *smaller than the digit one*, zeros after a decimal point that come before nonzero digits *are not* significant and serve only as place holders. For example, 0.0023 has two significant figures, 2 and 3. Zeros alone after a decimal point or zeros after a nonzero digit indicate a measurement, however, so these zeros *are* significant. The figure 0.00230, for example, has three significant figures since the 230 means 230 and not 229 or 231. Likewise, the figure 3.000 cm has four significant figures because the presence of the three zeros means that the measurement was actually 3.000 and not 2.999 or 3.001.

Multiplication and Division

When multiplying or dividing measurement figures, the answer may have no more significant figures than the *least* number of significant figures in the figures being multiplied or divided. This simply means that an answer can be no more accurate than the least accurate measurement entering into the calculation, and that you cannot improve the accuracy of a measurement by doing a calculation. For example, in multiplying 54.2 mi/hr × 4.0 hours to find out the total distance traveled, the first figure (54.2) has three significant figures but the second (4.0) has only two significant figures. The answer can contain only two significant figures since this is the weakest number of those involved in the calculation. The correct answer is therefore 220 miles, not 216.8 miles. This may seem strange since multiplying the two numbers together gives the answer of 216.8 miles. This answer, however, means a greater accuracy than is possible and the accuracy cannot be improved over the weakest number involved in the calculation. Since the weakest number (4.0) has only two significant figures the answer must also have only two significant figures, which is 220 miles.

The result of a calculation is **rounded** to have the same least number of significant figures as the least number of a measurement involved in the calculation. When rounding numbers, the last significant figure is increased by one if the number after it is five or larger. If the number after the last significant figure is four or less, the nonsignificant figures are simply dropped. Thus, if two significant figures are called for in the answer of the above example, 216.8 is rounded up to 220 because the last number after the two significant figures is 6, a number larger than 5. If the calculation result had been 214.8, the rounded number would be 210 miles.

Note that *measurement figures* are the only figures involved in the number of significant figures in the answer. Numbers that are **counted or defined** are not included in the determination of significant figures in an answer. For example, when dividing by 2 to find an average property of two objects, the 2 is ignored when considering the number of significant figures. Defined numbers are defined exactly and are not used in significant figures. For example, that a diameter is 2 times the radius is not a measurement. In addition, 1 kilogram is *defined* to be exactly 1000 grams so such a conversion is not a measurement.

Addition and Subtraction

Addition and subtraction operations involving measurements, as with multiplication and division, cannot result in an answer that implies greater accuracy than the measurements had before the calculation. Recall that the last digit in a measurement is considered to be uncertain because it is the result of an estimate. The answer to an addition or subtraction calculation can have this uncertain number *no farther from the decimal place than it was in the weakest number involved in the calculation*. Thus when 8.4 is added to 4.926, the weakest number is 8.4 and the uncertain number is .4, one place to the right of the decimal. The sum of 13.326 is therefore rounded to 13.3, reflecting the placement of this weakest doubtful figure.

Example Problem

In appendix III, "Experimental Error," an example was given of an experimental result of 511 Hz and an accepted value of 522 Hz, resulting in a calculation of

$$\frac{|522\ \text{Hz} - 511\ \text{Hz}|}{522\ \text{Hz}} \times 100\% = 2.1\%$$

Since 522 – 511 is 11, the least number of significant figures of measurements involved in this calculation is *two*. Note that the 100 does not enter into the determination since it is not a measurement number. The calculated result (from a calculator) is 2.1072797, which is rounded off to have only two significant figures, so the answer is recorded as 2.1%.

Appendix V: Conversion of Units

The measurement of most properties results in both a numerical value and a unit. The statement that a glass contains 50 cm³ of a liquid conveys two important concepts — the numerical value of 50, and the reference unit of cubic centimeters. Both the numerical value and the unit are necessary to communicate correctly the volume of the liquid.

When working with calculations involving measurement units, *both* the numerical value and the units are treated mathematically. As in other mathematical operations, there are general rules to follow.

Rule 1. Only properties with *like units* may be added or subtracted. It should be obvious that adding quantities such as 5 dollars and 10 dimes is meaningless. You must first convert to like units before adding or subtracting.

Rule 2. Like or unlike units may be multiplied or divided and treated in the same manner as numbers. You have used this rule when dealing with area (length × length = length², for example, or cm × cm = cm²) and when dealing with volume (length × length × length = length³, for example, or cm × cm × cm = cm³).

You can use the above two rules to create a **conversion ratio** that will help you change one unit to another. Suppose you need to convert 2.3 kilograms to grams. First, write the relationship between kilograms and grams:

$$1000 \text{ grams} = 1.000 \text{ kg.}$$

Next, divide both sides by what you wish to convert *from* (kilograms in this example):

$$\frac{1000 \text{ g}}{1.000 \text{ kg}} = \frac{1.000 \text{ kg}}{1.000 \text{ kg}}$$

One kilogram divided by one kilogram equals 1, just as 10 divided by 10 equals 1. Therefore, the right side of the relationship becomes 1:

$$\frac{1000 \text{ g}}{1.000 \text{ kg}} = 1$$

The 1 is usually understood — that is, not stated — and the operation is called *canceling*. Canceling leaves you with the fraction 1000 g/1.000 kg, which is a conversion ratio that can be used to convert from kg to g. You simply multiply the conversion ratio by the numerical value and unit you wish to convert:

$$2.3 \text{ kg} \times \frac{1000 \text{ g}}{1.000 \text{ kg}} = 2300 \text{ g}$$

The kg units cancel. Showing the whole operation with units only, you can see how you end up with the correct unit of g:

$$\text{kg} \times \frac{\text{g}}{\text{kg}} = \frac{\text{kg} \cdot \text{g}}{\text{kg}} = \text{g}$$

Since you did obtain the correct unit, you know that you used the correct conversion ratio. If you had blundered and used an inverted conversion ratio, you would obtain:

$$2.3 \times \frac{1.000 \text{ kg}}{1000 \text{ g}} = 23 \frac{\text{kg}^2}{\text{g}},$$

which yields the meaningless, incorrect units of kg^2/g. Carrying out the mathematical operations on the numbers and the units will always tell you if you used the correct conversion ratio or not.

Example Problem

A distance is reported as 100.0 km and you want to know how far this is in miles.

Solution

First, you need to obtain a conversion factor from a textbook or reference book, which usually groups similar conversion factors in a table. Such a table will show two conversion factors for kilometers and miles: (a) 1.000 km = 0.621 mi, and (b) 1.000 mi = 1.609 km. You select the factor that is in the same form as your problem. For example, your problem is 100.0 km = ? mi. The conversion factor in this form is 1.000 km = 0.621 mi.

Second, you convert this conversion factor into a **conversion ratio** by dividing the factor by what you want to convert *from*.

Conversion factor:	1.000 km = 0.621 mi
Divide factor by what you want to convert from:	$\dfrac{1.000 \text{ km}}{1.000 \text{ km}} = \dfrac{0.621 \text{ mi}}{1.000 \text{ km}}$
Resulting conversion ratio:	$\dfrac{0.621 \text{ mi}}{\text{km}}$

The conversion ratio is now multiplied by the numerical value and unit you wish to convert.

$$100.0 \text{ km} \times \frac{0.621 \text{ mi}}{\text{km}}$$

$$100.0 \times 0.621 \frac{\text{km} \cdot \text{mi}}{\text{km}}$$

$$62.1 \text{ miles}.$$

Appendix VI: Scientific Notation

Most of the properties of things that you might measure in your everyday world can be expressed with a small range of numerical values together with some standard unit of measure. The range of numerical values for most everyday things can be dealt with by using units (1's), tens (10's), hundreds (100's), or perhaps thousands (1,000's). But the universe contains some objects of incredibly large size that require some very big numbers to describe. The sun, for example, has a mass of about 1,970,000,000,000,000,000,000,000,000,000 kg. On the other hand, very small numbers are needed to measure the size and parts of an atom. The radius of a hydrogen atom, for example, is about 0.00000000005 m. Such extremely large and small numbers are cumbersome and awkward since there are so many zeros to keep track of, even if you are successful in carefully counting all the zeros. A method does exist to deal with extremely large or small numbers in a more condensed form. The method is called **scientific notation**, but it is also sometimes called *powers of ten* or *exponential notation* since it is based on exponents of 10. Whatever it is called, the method is a compact way of dealing with numbers that not only helps you keep track of zeros but also provides a simplified way to make calculations as well.

In algebra you save a lot of time (as well as paper) by writing (a × a × a × a × a) as a^5. The small number written to the right and above a letter or number is a superscript called an **exponent**. The exponent means that the letter or number is to be multiplied by itself that many times. For example, a^5 means "a" multiplied by itself five times, or a × a × a × a × a. As you can see, it is much easier to write the exponential form of this operation than it is to write out the long form.

Scientific notation uses an exponent to indicate the power of the base 10. The exponent tells how many times the base, 10, is multiplied by itself. For example:

$$10,000. = 10^4$$

$$1,000. = 10^3$$

$$100. = 10^2$$

$$10. = 10^1$$

$$1. = 10^0$$

$$0.1 = 10^{-1}$$

$$0.01 = 10^{-2}$$

$$0.001 = 10^{-3}$$

$$0.0001 = 10^{-4}$$

This table could be extended indefinitely, but this somewhat shorter version will give you an idea of how the method works. The symbol 10^4 is read as "ten to the fourth power" and means $10 \times 10 \times 10 \times 10$. Ten times itself four times is 10,000, so 10^4 is the scientific notation for 10,000. It is also equal to the number of zeros between the 1 and the decimal point. That is, to write the longer form of 10^4 you simply write 1, then move the decimal point four places to the *right*; hence ten to the fourth power is 10,000.

The powers of ten table also shows that numbers smaller than one have negative exponents. A negative exponent means a reciprocal:

$$10^{-1} = \frac{1}{10} = 0.1$$

$$10^{-2} = \frac{1}{100} = 0.01$$

$$10^{-3} = \frac{1}{1000} = 0.001$$

To write the longer form of 10^{-4}, you simply write 1 then move the decimal point four places to the *left*; hence ten to the negative fourth power is 0.0001.

Scientific notation usually, but not always, is expressed as the product of two numbers: (1) a number between 1 and 10 that is called the **coefficient**, and (2) a power of ten that is called the **exponential**. For example, the mass of the sun that was given in long form earlier is expressed in scientific notation as

$$1.97 \times 10^{30} \text{ kg}$$

and the radius of a hydrogen atom is

$$5.0 \times 10^{-11} \text{ m.}$$

In these expressions, the coefficients are 1.97 and 5.0 and the power of ten notations are the exponentials. Note that in both of these examples, the exponential tells you where to place the decimal point if you wish to write the number all the way out in the long form. Sometimes scientific notation is written without a coefficient, showing only the exponential. In these cases the coefficient of 1.0 is understood; that is, not stated. If you try to enter a scientific notation in your calculator, however, you will need to enter the understood 1.0 or the calculator will not be able to function correctly. Note also that 1.97×10^{30} kg and the expressions 0.197×10^{31} kg and 19.7×10^{29} kg are all correct expressions of the mass of the sun. By convention, however, you will use the form that has one digit to the left of the decimal.

Example Problem

What is 26,000,000 in scientific notation?

Solution

Count how many times you must shift the decimal point until one digit remains to the left of the decimal point. For numbers larger than the digit 1, the number of shifts tells you how much the exponent is increased, so the answer is 2.6×10^7, which means the coefficient 2.6 is multiplied by 10 seven times.

Example

What is 0.000732 in scientific notation? (Answer: 7.32×10^{-4}.)

Multiplication and Division

It was stated earlier that scientific notation provides a compact way of dealing with very large or very small numbers but provides a simplified way to make calculations as well. There are a few mathematical rules that will describe how the use of scientific notation simplifies these calculations.

To *multiply* two scientific notation numbers, the coefficients are multiplied as usual and the exponents are *added* algebraically. For example, to multiply (2×10^2) by (3×10^3), first separate the coefficients from the exponentials,

$$(2 \times 3) \times (10^2 \times 10^3),$$

then multiply the coefficients and add the exponents,

$$6 \times 10^{(2+3)} = 6 \times 10^5.$$

Adding the exponents is possible because 10^2 3 10^3 means the same thing as $(10 \times 10) \times (10 \times 10 \times 10)$, which equals $(100) \times (1,000)$, or 100,000, which is expressed as 10^5 in scientific notation. Note that two negative exponents add algebraically, for example, $10^{-2} \times 10^{-3} = 10^{[(-2) + (-3)]} = 10^{-5}$. A negative and a positive exponent also add algebraically, as in $10^5 \times 10^{-3} = 10^{[(+5) + (-3)]} = 10^2$.

If the result of a calculation involving two scientific notation numbers does not have the conventional one digit to the left of the decimal, move the decimal point so it does, changing the exponent according to which way and how much the decimal point is moved. Note that the exponent increases by one number for each decimal point moved to the left. Likewise, the exponent decreases by one number for each decimal point moved to the right. For example, $938. \times 10^3$ becomes 9.38×10^5 when the decimal point is moved two places to the left.

To *divide* two scientific notation numbers, the coefficients are divided as usual and the exponents are *subtracted*. For example, to divide (6×10^6) by (3×10^2), first separate the coefficients

411

from the exponentials,

$$\left(\frac{6}{3}\right) \times \left(\frac{10^6}{10^2}\right)$$

then divide the coefficients and subtract the exponents,

$$2 \times 10^{(6-2)} = 2 \times 10^4$$

Note that when you subtract a negative exponent, for example, $10^{[(3)-(-2)]}$, you change the sign and add, $10^{(3+2)} = 10^5$.

Example Problem

Solve the following problem concerning scientific notation:

$$\frac{\left(2 \times 10^4\right) \times \left(8 \times 10^{-6}\right)}{8 \times 10^4}$$

Solution

First, separate the coefficients from the exponentials,

$$\frac{2 \times 8}{8} \times \frac{10^4 \times 10^{-6}}{10^4},$$

then multiply and divide the coefficients and add and subtract the exponents as the problem requires,

$$2 \times 10^{\{[(4)+(-6)]-[4]\}}$$

Solving these remaining operations gives 2×10^{-6}.

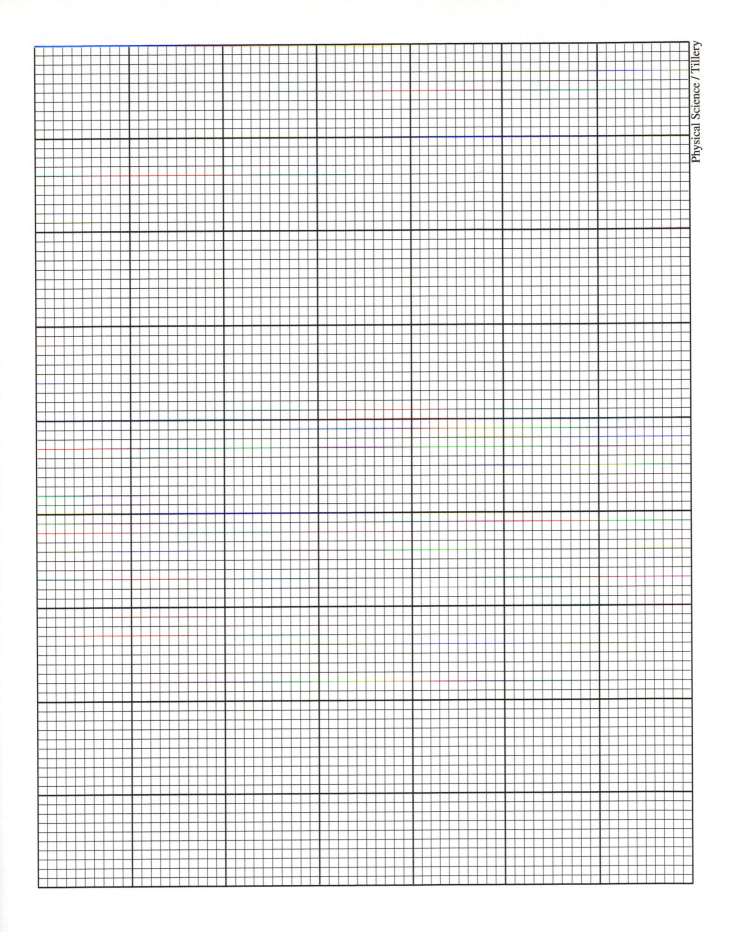

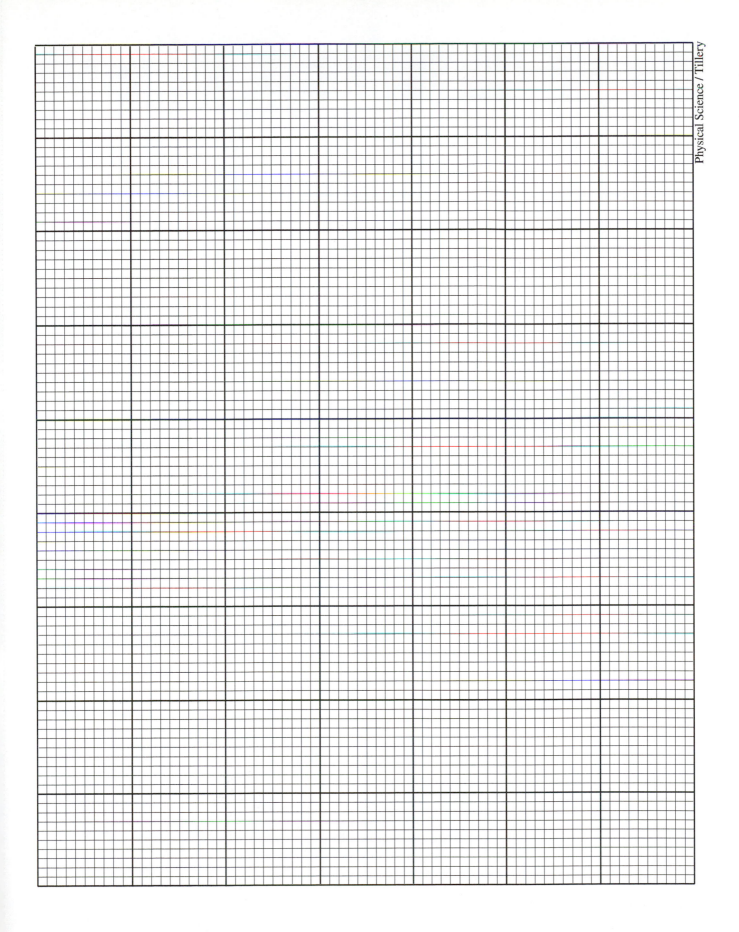

Notes

Notes

Relative Humidity Chart

Dry-bulb ↓ (°C)	Difference between wet-bulb and dry-bulb temperatures (°C)																		
	1	2	3	4	5	6	7	8	9	10	11	12	13	14	15	16	17	18	19
0	81	64	46	29	13														
1	83	66	49	33	17														
2	84	68	52	37	22	7													
3	84	70	55	40	26	12													
4	86	71	57	43	29	16													
5	86	72	58	45	33	20	7												
6	86	73	60	48	35	24	11												
7	87	74	62	50	38	26	15												
8	87	75	63	51	40	29	19	8											
9	88	76	64	53	42	32	22	12											
10	88	77	66	55	44	34	24	15	6										
11	89	78	67	56	46	36	27	18	9										
12	89	78	68	58	48	39	29	21	12										
13	89	79	69	59	50	41	32	23	15	7									
14	90	79	70	60	51	42	34	26	18	10									
15	90	80	71	61	53	44	36	27	20	13	6								
16	90	81	71	63	54	46	38	30	23	15	8								
17	90	81	72	64	55	47	40	32	25	18	11								
18	91	82	73	65	57	49	41	34	27	20	14	7							
19	91	82	74	65	58	50	43	36	29	22	16	10							
20	91	83	74	66	59	51	44	37	31	24	18	12	6						
21	91	83	75	67	60	53	46	39	32	26	20	14	9						
22	92	83	76	68	61	54	47	40	34	28	22	17	11	6					
23	92	84	76	69	62	55	48	42	36	30	24	19	13	8					
24	92	84	77	69	62	56	49	43	37	31	26	20	15	10	5				
25	92	84	77	70	63	57	50	44	39	33	28	22	17	12	8				
26	92	85	78	71	64	58	51	46	40	34	29	24	19	14	10	5			
27	92	85	78	71	65	58	52	47	41	36	31	26	21	16	12	7			
28	93	85	78	72	65	59	53	48	42	37	32	27	22	18	13	9	5		
29	93	86	79	72	66	60	54	49	43	38	33	28	24	19	15	11	7		
30	93	86	79	73	67	61	55	50	44	39	35	30	25	21	17	13	9	5	
31	93	86	80	73	67	61	56	51	45	40	36	31	27	22	18	14	11	7	
32	93	86	80	74	68	62	57	51	46	41	37	32	28	24	20	16	12	9	5
33	93	87	80	74	68	63	57	52	47	42	38	33	29	25	21	17	14	10	7
34	93	87	81	75	69	63	58	53	48	43	39	35	30	28	23	19	15	12	8